Commercial
Management
in Construction

Commercial Management in Construction

Ian Walker
Bowey Construction

and

Robert Wilkie
University of Northumbria at Newcastle

Blackwell
Science

© 2002 by Blackwell Science Ltd,
a Blackwell Publishing Company
Editorial Offices:
Osney Mead, Oxford OX2 0EL, UK
   Tel: +44 (0)1865 206206
Blackwell Science, Inc., 350 Main Street,
Malden, MA 02148-5018, USA
   Tel: +1 781 388 8250
Iowa State Press, a Blackwell Publishing
Company, 2121 State Avenue, Ames,
Iowa 50014-8300, USA
   Tel: +1 515 292 0140
Blackwell Publishing Asia Pty Ltd,
550 Swanston Street, Carlton South, Melbourne,
Victoria 3053, Australia
   Tel: +61 (0)3 9347 0300
Blackwell Wissenschafts Verlag, Kurfürstendamm
57, 10707 Berlin, Germany
   Tel: +49 (0)30 32 79 060

The right of the Author to be identified as the
Author of this Work has been asserted in
accordance with the Copyright, Designs and
Patents Act 1988.

All rights reserved. No part of this publication
may be reproduced, stored in a retrieval system,
or transmitted, in any form or by any means,
electronic, mechanical, photocopying, recording
or otherwise, except as permitted by the UK
Copyright, Designs and Patents Act 1988,
without the prior permission of the publisher.

First published by Blackwell Science, 2002

Library of Congress
Cataloging-in-Publication Data
is available

ISBN 0-632-05827-7

A catalogue record for this title is available from
the British Library

Produced and typeset by Gray Publishing,
Tunbridge Wells, Kent
Printed and bound in Great Britain by
MPG Books Ltd, Bodmin, Cornwall

For further information on
Blackwell Science, visit our website:
www.blackwell-science.com

Contents

	Foreword	vii
	Acknowledgements	viii
Chapter 1	Introduction	1
Chapter 2	Budgets and Forecasting	6
Chapter 3	Interim Valuations	17
Chapter 4	Subcontractors	57
Chapter 5	Cost Value Comparisons	78
Chapter 6	Contracts, Certificates and Notices	93
Chapter 7	Teamwork and Partnering	109
Appendix 1	A worked example	119
Appendix 2	Cost value comparison: an easy guide	167
	Bibliography	173
	Glossary of Terms	174
	Index	177

I would like to dedicate this book to three very important people in my life, sadly they all died too early and will always be greatly missed.

To Noel (my dad), Bob (my uncle) and Jack (my father-in-law): thanks for the time and the model.

Ian Walker

Foreword

This book is probably unique in having as its focus the financial management of the on-site construction process. The role of the contractor's quantity surveyor has an interesting history – one which probably pre-dates that of the so-called professional quantity surveyor (PQS). More recently, while the role of the PQS in contractual matters has been increasingly marginalised by new approaches to construction procurement, that of the contractor's quantity surveyor has developed into a distinctive concern for commercial management in the construction industry.

The publication of this book is therefore timely. It describes a role founded on skill in the quantification and valuation of construction, but now extended into a comprehensive responsibility for management accounting and financial information systems. The text is presented as a practical manual, providing worked examples of a wide range of techniques. The book also looks forward into the role of the commercial manager in the new world of partnering and key performance indicators – a world in which best value procurement and continuous improvement are replacing the outdated culture of lowest price selection and contractual claims.

I have no doubt that it will be of value to both students and practitioners.

Professor John Bale
Past President of the Chartered Institute of Building
Willmott Dixon Professor of Construction Management, Leeds Metropolitan University

Acknowledgements

There can be no doubt that without the help and encouragement of many friends and colleagues this publication would not have been completed. Their support and input have proved invaluable and I sincerely thank them all. First, thanks to my co-author, Bob Wilkie, without whom I would not have started, and to his colleagues from the University of Northumbria at Newcastle, Simon Murray and Phil Thomas, for their contributions and encouragement; to Andy Ross of Liverpool John Moores University for his help and encouragement; and to all my friends and colleagues at my place of work, in particular to David Wright for proof-reading completed sections.

I am grateful to Roddy Gordon and Andrea Gardner of Watson Burton Solicitors for keeping me on the right track with contract matters. Thanks also to the Sir William Turners team of Carol Middleton of Tees Valley Housing, John Wade and Doug Whelan of ADG Architects, David Walker and Justine Platt of Faithfull and Gould and James Walker, John Bramley and Jack Downey of Bowey Construction for allowing the use of their KPIs and their partnering benchmark targets.

To my family, my wife Marilyn and son James Grant and daughter Katherine Jill, and I must not forget my two dogs, Sam and Hannah: thank you all for your help, understanding and forbearance; perhaps now we can get back to normal!

Finally, thanks to two people who have been there when needed and who have given continual support during the various stages of producing this book: Phyl Lawrence of the Chartered Institute of Building helped me through the original thesis which was the forerunner to this book; and Madeleine Metcalfe of Blackwell Publishing kept me going over the past two years or so, and without her encouragement Bob and I would still be writing.

Copyright Permissions

The authors and publishers would like to thank the following for their kind permission to use their material:

Extracts from JCT forms of Contract and Interim Certificates by kind permission of Joint Contracts Tribunal Ltd and RIBA Companies Ltd.

Extracts from 'Diligently and Faithfully' by kind permission of the Chartered Institute of Building.

Extracts from the 'Definition of Prime Cost of Daywork' by kind permission of the National Federation of Builders.

Examples from computerised estimating and valuation systems by kind permission of Conquest Ltd.

Extracts from Construction Industry Scheme by kind permission of Crown Copyright.

Extracts from VAT Building and Construction, Notice 708 August 1997, by kind permission of HM Customs and Excise.

1 Introduction

This book is intended to help commercial managers and contractors' quantity surveyors with the requirements of life in a contractor's office at an early stage in their careers. The specialist functions and processes that they will be required to perform can be considerably different from the general studies they may undertake during a typical formal full-time quantity surveying course. In addition to the benefits that it will bring to those directly involved with the commercial management of a construction company, the book should give an insight into the requirements of the contractor's quantity surveyor for those outside the contracting environment, but still within the quantity surveying discipline. If each team member understands the needs of the other, it will promote more harmonious relationships.

The text does not reflect the only way to approach or undertake the contractor's quantity surveyor's work. It is an account of the commercial aspects required and an insight into the day-to-day work completed by the contractor's quantity surveyor. There are many examples of how to achieve results, together with general background principles. The text does not purport to be definitive on legal or contractual requirements. Wallace (1984), Keating (1978) and Murdoch and Hughes (1996) deal very effectively with these matters. Similarly, it is not the intention to describe in detail the general principles or requirements of measurement; Seeley (1988) and Willis *et al.* (1998) cover these areas. Where possible the text has followed the general format of other publications so that comparisons can be drawn between the roles of contractors' quantity surveyors and quantity surveyors in private practice.

While building companies differ in size and structure, or what they require from their quantity surveyors, it is hoped that what follows demonstrates sufficiently the services management may require from the quantity surveyor and how best to achieve these goals to assist in the economic and successful running of the company.

Appendix 1 contains an example showing a fully worked through interim valuation. This will take the student from the initial interim valuation submission made by the contractor's quantity surveyor, through to the completed cost valuation comparison. The text also provides examples of a subcontract comparison both for the purpose of placing a subcontract order and for use with the subcontract liability for inclusion within the cost value comparison.

Chapter 2, *Budgets and Forecasting*, and Chapter 5, *Cost Value Comparisons*, cover the requirements of the statement of standard accounting practice no. 9 (SSAP9), highlighting the need to produce both management and financial accounts, describing the differences between the two sets of figures and explaining the methodology used in the completion of these accounts.

As well as the technical skills required of those dealing with the commercial side of a contractor's business, Chapter 7 deals with teamwork and, in particular, the current methodology of working in partnership. While the chapter contains, details of techniques that can be used and to assist those involved in this market sector of the industry, the main thrust is that of *teamwork* and the benefits that can be gathered by using all the resources available within the whole of the supply chain. Working in unison towards a common goal, and to mutual benefit of all parties, will produce better results for all those involved whichever procurement method is used.

The construction industry is going through major changes in its methodology of procurement and working practices. All elements of the industry are having to re-examine their roles and retrain where necessary.

As the text will describe, there are many new procurement routes to consider. Partnering requires in many instances a whole culture change within the industry; there is a need for the rebuilding of trust between parties, for more transparent book-keeping and for teams to work together in a no-blame culture, away from confrontation.

Commercial managers need to be aware of and use new techniques, to keep abreast of new technology, to set targets and measure themselves against these targets to achieve the continuous improvements required to keep them at the forefront of the industry.

History

Before moving on to the main text of the book we will look at how the various disciplines came about by looking at their history and how they formed into the institutions we know today.

The Royal Institution of Chartered Surveyors (RICS)

The history of the RICS shows there was a Reading-based firm of quantity surveyors in 1785. Both Seeley (1988, 1997) and Willis *et al.* (1998) set out how the quantity surveying profession evolved, and there is evidence that there were other firms around at that time. The first method of measurement was produced following a meeting of Scottish quantity surveying firms in 1902, before which craftsmen employed measurers to deal with their accounts. During the industrial revolution it was customary for master builders and general contractors to undertake major sections of work on behalf of architects, who at that time were responsible for both the construction and design of the projects. They often worked from minimal information with much of the works being designed as the project proceeded. Costings were only prepared by the measurers and submitted to the client for payment at the end of a project. At this juncture materials scheduled on invoices were checked against works measured on site. In some cases extravagant claims for waste of material were made by the master craftsmen and general contractors, which led to disputes with the architects who, in turn, would employ quantity surveyors to contest the claims of the contractor's quantity surveyors or measurers and agree the final accounts.

The origin of the quantity surveyor goes back a stage further. Initially, as indicated above, the architect would employ craftsmen to carry out sections of work, but after the fire of London these master craftsmen began to form craft firms which would complete larger sections of work. Architects were inundated with work and could not continue to carry out the whole process of design and costing. In addition, where the craft firms were invited to tender for work by the architect, they often employed the services of measurers to prepare schedules of work to be carried out. Employers and architects soon realised that each of the craft firms were employing their own measurers to produce their tender costing, a situation that was, as a result of the additional overhead costs, leading to increases in tender prices.

During the late nineteenth century, as buildings became more complex, the practice of contractors employing their own measurers was revised. Contractors began to appoint one measurer/surveyor to prepare quantities or work schedules on their behalf, with the winning tenderer paying the account to the measurer/surveyor. This produced not only lower costs to the builders but also uniformity of documentation for each builder to work from.

These measurers/quantity surveyors would work in conjunction with the architect's quantity surveyor to produce work schedules and bills of quantities. Latterly, the role became more of a separate practice, with this scenario being the predecessor of the independent quantity surveyor we know today.

The Institution of Surveyors (later known as the Surveyors' Institution) was founded in 1863 with a membership of 200, some 74 years after the establishment of the first practice. Queen Victoria granted the Institution a Royal Charter in 1881, by which time the membership had grown to 500. By 1918 the membership stood at approximately 5000. In 1922 the Quantity Surveyors' Association amalgamated with the Institution and the name was altered to the Chartered Surveyors' Institution in 1930. In 1946 King George VI, who was then patron of the Institution, commanded that the Institution should in future be known as the Royal Institution of Chartered Surveyors. For many years membership of the Institution was restricted to principals or assistants in private practice, or officials of central and local government departments who had passed the Institution's examinations, which were first held in 1881. Surveyors employed by contracting and other commercial organisations were not admitted for membership, nor were they permitted to take the Institution's examinations. Membership of RICSs now stands at 105 000 members, with the quantity surveying section accounting for approximately 24 000 of the membership.

In comparison with other institutes, the Institute of Quantity Surveyors was founded relatively recently, in 1938. The main difference between this and the RICS was that the Institute of Quantity Surveyors would admit members from contracting organisations as well as quantity surveyors engaged in private practice. The Institute of Quantity Surveyors was incorporated into RICS in 1983, thus drawing many of those employed in the quantity surveying discipline within the construction industry into one institution.

The Chartered Institute of Building (CIOB)

Many contractor's quantity surveyors select the CIOB as their professional body. The Institute's short history 'Diligently and Faithfully' (Powell, 1997) describes how the Institute was founded. The Builders' Society was formed in 1834 by a group of leading London builders whose primary concerns were the industrial unrest during the Industrial Revolution coupled with new government legislation. For many years the society numbers remained small, starting with only 17 members, all respected and established figures in the industry. Their collective objective was to foster:

> the promotion of friendly intercourse, the interchange of useful information with increasing uniformity, and respectability in the conduct of business.

Membership continued to grow at a relatively slow pace and comprised some 75 builders around 1870. The difference between the Builders' Society and other institutions such as RICS and the Royal Institute of British Architects (RIBA) was that the builders employed many thousands within their companies. Following a strike in 1859 at the contractors Trollope and a general meeting of the trade in London, a new employers' association, the Central Association of Master Builders (CAMB), was formed. The founding of this new association eventually led to the Builders' Society ceasing to play the role of an employers' association.

In 1868 the Builders' Society produced a report on terms of contract which led to discussions with RIBA. These discussions came to fruition in 1872 with the introduction of a simple standard form of contract. Despite this movement the society was not flourishing, membership dropped to only about 40 elected members and meetings were changed from monthly to quarterly.

In June 1884, some 50 years after its formation, the Builders' Society was replaced by the Institute of Builders, with its first President being Colonel Stanley Bird. For many years membership was in effect restricted to principals of large firms. It was not until 1923, when membership stood at 402 and after many years of preparation, that examinations became the method of entry into the Institute. As the Institute passed its centenary, membership grew to 1200, with members admitted by examination now exceeding non-examination members. Heavily influenced by Albert Costain and Peter Shepherd, the Institute's stature continued to grow, with membership standing at 5000 in 1963. The structure of examinations changed, with more emphasis being placed on building management.

In 1965 the Institute was renamed the Institute of Building, the name change reflecting the broader vision of what it stood for. The Institute moved to its own premises at Englemere in 1972, received a Royal Charter in September 1980 and began the new millennium with a membership of around 36 500.

The Royal Institute of British Architects (RIBA)

RIBA was formed out of several societies dating back to the 1790s, the Architects' Club and the London Architectural Society being two of the more prominent societies at the time. RIBA was established in early 1834 as the Society of British Architects, became the Institute of British Architects later that year and received its first Royal Charter in 1837. In March 1835 Lord De Gray was elected President. Members of the Institute were elected by ballot, with only architects of unimpeachable character and evident respectability being encouraged to join.

Training for architects during this period was generally unsatisfactory, as articled pupils of practising architects. In 1863 the Institute appointed a board of examiners and although examinations were voluntary candidates needed to succeed in them before election as an Associate of the Institute. In 1876 a special committee recommended that all associates elected after May 1882 should have passed a professional examination. Shortly after the introduction of the examination process the Institute introduced the annual award of the Royal Gold Medal for Architecture.

For many years the Institute remained a London-based organisation, with other alliances from the provinces combining to form the Architectural Alliance, but in 1876 it was recommended that closer relationships with the provinces be developed. The first of these alliances was made in 1889 and by 1939 a national and commonwealth network of allied societies was in place. Between this early period and 1906 there was great debate within the RIBA and breakaway bodies such as the Society of Architects who were campaigning for statutory registration, which would exclude those who had not passed the professional examinations. Those against these proposals, the 'memorialists', thought it impossible to examine imagination, power of design and refinement of taste and judgement. They feared that the system would benefit the profession of architects and not the art of architecture, and campaigned vigorously to change the emphasis of the Institute from architect to architecture.

In 1931 an act was passed that established an Architects' Registration Council of the United Kingdom as an independent body and by 1938 the use of the title of 'architect' was restricted to those on this register.

RIBA remains a totally autonomous and self-governing body with a membership of some 30 000 architects.

The National Federation of Builders

The National Federation of Builders is the largest trade association for employers in the construction industry. Established in 1872, the Federation has considerable experience in supporting its members and provides a wide range of support to its members, including such matters as employment affairs, health and safety issues, legal and contractual matters, technical advice, European legislation and tax affairs. The National Federation of Builders has a membership of 3500 companies throughout England and Wales.

Recommended Further Reading

Keating, D. (1978) *Building Contracts*. Sweet and Maxwell.
Murdoch, J. and Hughes, W. (1996) *Construction Contracts Law and Management*. E&FN Spon.
Powell C. (1997) *Diligently and Faithfully. A Short History of The Chartered Institute of Building*. Chartered Institute of Building.
Seeley, I.H. (1988) *Building Quantities Explained*. Macmillan.
Seeley, I.H. (1997) *Quantity Surveying Practice*. Macmillan.
Wallace, I.N.D. (1984) *Hudson's Building Contracts*. Sweet and Maxwell.
Willis, C.J., Ashworth, A. and Willis, J.A. (1994) *Practice and Procedure for the Quantity Surveyor*. Blackwell Science.
Willis, A. and Trench, W. (1998) *Willis's Elements of Quantity Surveying*. Blackwell Science.

2 Budgets and Forecasting

A *budget* is described in the Oxford Dictionary as an *annual estimate*. In contracting terms, this generally relates to an assessment of turnover or net cost of production (NCP), an assessment of overhead and finally an assessment of profitability. All companies need to know how they expect to perform over the financial year. In addition, they must monitor their performance by comparing actual results against the original budget figures.

The completion of the contractor's annual budget will conventionally be within the remit of and the responsibility of the finance director. However, to compile the overall company budget, it is necessary to start with the lowest or most subordinate elements within the organisation. In the case of a contracting organisation this will mean analysing individual projects in terms of their expenditure and performance. This assessment of each project will be carried out by the contractor's quantity surveyor.

This assessment is not restricted to projects in hand. To arrive at an overall picture, it is necessary to consider all aspects of work that may be carried out during any particular financial year. Ongoing projects are only part of the equation. The analysis can be divided into four main headings:

- current projects
- probable projects
- possible projects
- anticipated tender conversions.

In general, these four sections will cover most eventualities within the contractor's estimate of performance within the financial year. Taking each element in turn consideration should be given to the following factors.

Current Projects

This element relates to two types of project: first, projects that are already current or work in progress but that will not be completed within the current financial year, and secondly, projects for which a contract has been agreed and signed by both contracting parties, but that will not start until during the financial year under review.

Probable Projects

To qualify for inclusion in this section a project has to be 'almost there'. This may relate to projects being negotiated at the time, or projects where the contractor's tender has been accepted but no formal order or contract is yet in place.

Possible Projects

To establish any budget there is a clear need to look ahead, to assess what might be. From a contractor's point of view it is necessary to attempt to assess what future projects they may be involved in. To complete this exercise the contractor's quantity surveyor will have to look ahead to future tendering commitments. Most contractors will monitor anticipated incoming tenders; they will know their client base and will know their tendering commitment some months in advance. Unsolicited tender enquiries may arrive at the contractor's door, but with a good marketing and estimating management system in place there should be some degree of certainty regarding which and when projects will be forthcoming. Within this element of the budget calculation the contractor's quantity surveyor must establish a schedule of anticipated projects, which of these are of particular interest to the contractor, and which the management team envisage they will be better placed to win in a tender situation.

The contractor may also be involved, albeit in the very early stages, in projects that may be secured by means other than in open tender. This could be by negotiation, or perhaps on a partnered basis. Such projects need to be scheduled and appraised accordingly for inclusion within this section of the budget analysis.

Anticipated Tender Conversions

This element needs to be completed, in part, by historical information recorded from previous years. There is a need to analyse tender success ratios – e.g. 1:4, 1:6 or 1:10 and, using the ratio figure, calculate the anticipated number of projects that the contractor might reasonably be awarded during the next financial year. Similarly, the turnover and profit elements are calculated from comparable historical records.

Within this section there is also a need to consider other elements of the contractor's business. Staffing levels need to be taken into account, both on and off site. The knowledge and experience of the staff also need to be analysed when assessing which types and values of project are best suited to the teams.

The contractor's quantity surveyor should also consider earning potential, whether there is a pattern of profitability, and which types of project are capable of higher earning potential than others. All such matters need to be considered during the compilation of budget assessments.

A further matter that may require consideration is whether the contractor has a planned expansion programme underway and, if so, what strategies have been put in place to increase incoming tenders and turnover. This information will greatly assist in the preparation of this element of the budget.

As part of this section, consideration should also be given to future work requirements, perhaps for the following year. Do the figures produced for the first three sections of the budget or forecast indicate a shortfall of work to span into the following year, or is there a need to obtain short projects to fill the current year's orderbook? Whichever situation applies, and it may be both, the contractor can produce an action plan to facilitate the requirements.

The total of these four sections should form the basis of the contractor's turnover and project profit expectations for the coming financial year.

Another section that may be considered is turnover and profit that may be generated from completed projects, but where final account work has not been completed. If this criterion is considered then it should form a further section in its own right. As with the completion of cost value comparisons detailed in Chapter 5, the contractor's quantity surveyor must err on the side of caution when drawing figures into this section.

The contractor's operation is subdivided for ease of production and analysis by others. As already indicated, it is necessary to take all aspects to their lowest denominator and build from there.

Clearly, the analysis of current and probable projects and that of projects in their final accounting stage will be easier and should be more accurate than those in the remaining elements, but all should be considered individually on their own merit. Each project must therefore be looked at in turn and all sections analysed thoroughly.

The starting point has to be the original pricing document or tender, i.e. the tender sum and the estimated profit margin at tender time. By using these figures from the original tender, the contractor will be able to monitor true performance as the projects proceed. These figures can be taken into account in future budgets or tender assessments.

Figure 2.1 highlights a suitable format for these assessments that can be used for budgets or general forecasting.

Once the initial tender figures have been established the original tender sum is adjusted back to the net cost by the deducting tender profit margin.

It is then necessary to adjust both sides of the equation for those items scheduled, or any other matters that the contractor's quantity surveyor considers require adjustment to either turnover or profit. All known facts should be assimilated, including any increase in profit from subcontract orders placed, and general material buying or errors that have been discovered in the original tender which increase or decrease the anticipated return, to produce an anticipated revised outcome for each project. The schedule highlighted should be used as the project summary sheet with back-up schedules being used in support of any figures used.

When analysing subcontract or material figures a full schedule should be produced to record the figures accurately. With regard to subcontractors, for instance, this schedule should include a list of the original subcontractors used in the estimate, together with their net costings (Figure 2.2). The difference in cost can then be compared with the final choice of individual subcontractors and the resultant figure can be transferred to the project forecast schedule (Figure 2.1). Material analysis can be completed using a similar technique and schedule.

Both labour and plant should also be reassessed: the management team should analyse the total project and produce both labour and plant histograms of resources to be used. These can then be assessed against the original tendered allowances and the resultant figures can be incorporated in whole or in part, as the team considers appropriate, on the project assessment. Computer-aided project management packages, such as Microplanner, can greatly assist in the preparation and dissemination of such figures.

Budgets and forecasting 9

Murribell Builders Ltd
Budget / Forecast

Project type ☐

CONTRACT

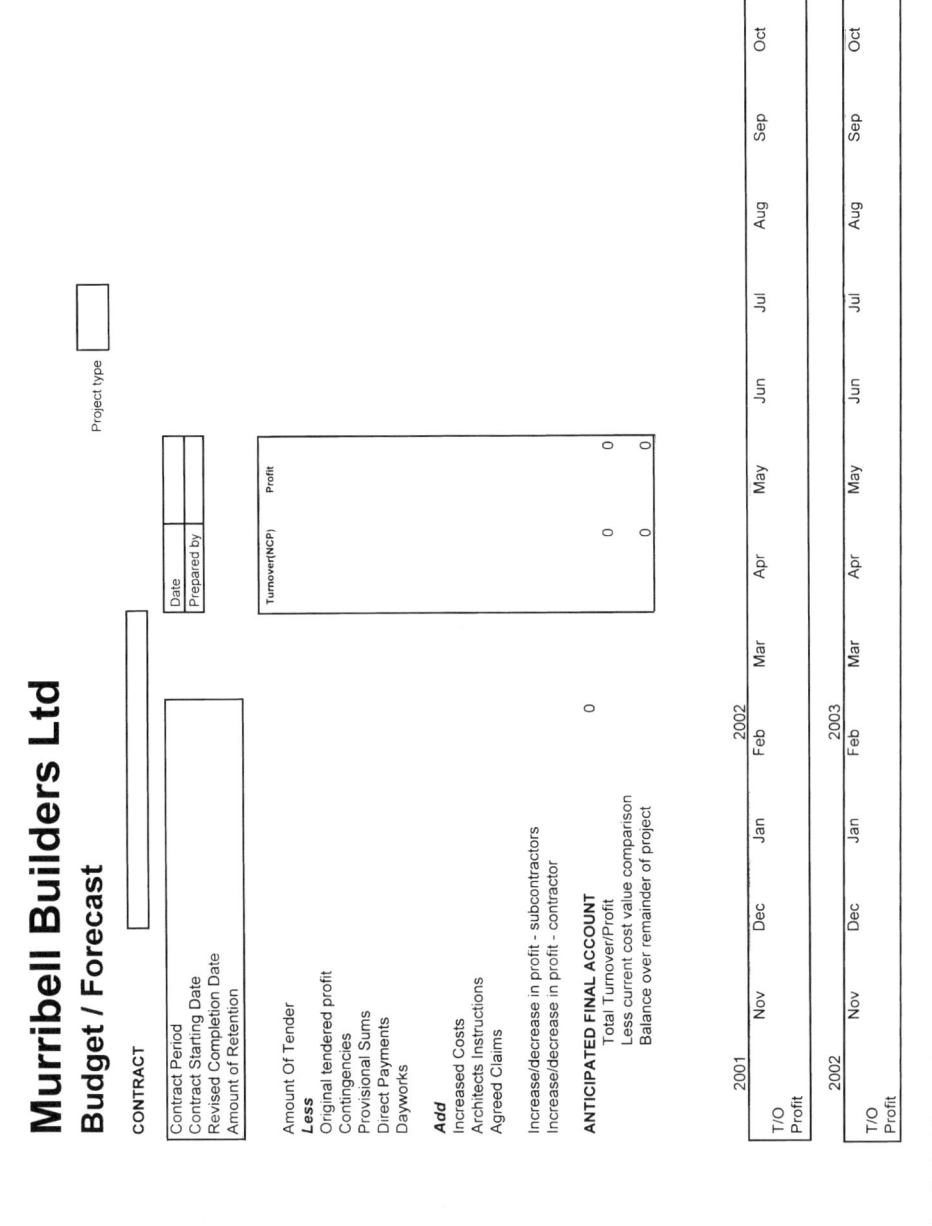

Date
Prepared by

	Turnover(NCP)	Profit

Contract Period
Contract Starting Date
Revised Completion Date
Amount of Retention

Amount Of Tender
Less
Original tendered profit
Contingencies
Provisional Sums
Direct Payments
Dayworks

Add
Increased Costs
Architects Instructions
Agreed Claims

Increase/decrease in profit - subcontractors
Increase/decrease in profit - contractor

ANTICIPATED FINAL ACCOUNT
   Total Turnover/Profit   0   0
   Less current cost value comparison   0
   Balance over remainder of project   0   0

2001

	Nov	Dec	Jan	Feb	Mar	Apr	May	Jun	Jul	Aug	Sep	Oct	Total
T/O													0
Profit													0

2002

2002

	Nov	Dec	Jan	Feb	Mar	Apr	May	Jun	Jul	Aug	Sep	Oct	Total
T/O													0
Profit													0

2003

Figure 2.1 Project forecast.

Murribell Builders Ltd

SUBCONTRACT ANALYSIS

Project				Contract No	
				Date	
Sub-Trade	Subcontractor in Tender	Net Tender Value	Selected Subcontractor	Net Accepted Tender	Saving/ Loss
Demolition					
Electrical					
Mechanical					
Lift Installation					
Plumbing					
Steelwork					
Roof Covering					
Felt Roofing					
Asphalt					
Plaster work					
Render					
Suspended Clg					
Dry Lining					
Fire Barriers					
Partitions					
Glazing					
Insulation					
Painter work					
Mastic Pointing					
Floor Covering					
Wall Tiling					
Fire Protection					
Landscaping					
Tarmac					
Road Markings					
Safety Fasteners					
Fencing					
P.C. Floors					
Secondary Glazing					
Handrails					
Scaffolding					
Totals					
			Average percentage gain		%

Figure 2.2 Subcontract comparison.

The analysis should not be restricted to the above matters; all areas need to be considered. Is there merit in early completion; can preliminary costs be saved? What influence can the contractor's team have on amending the specification; is there a benefit to the contractor here? Is there a significant element of provisional sums, which are yet to be expended, from which the contractor can expect a certain level of recovery of overhead and profit which will enhance profitability? Many questions need to be asked and many areas need to be considered by the contractor's team.

Consideration should be given not only to the positive side, but also to areas of risk. At the onset of any project the contractor's quantity surveyor would be well advised to work with the project team and produce a schedule of risk elements. These elements may be a risk either to the programme or to profitability; whichever is applicable, the anticipated outcome must be taken into consideration when assessing all projects.

Clearly, the more advanced the project is the more accurate the assessment should be, but even on very embryonic projects full consideration should be given to the final projected outcome. In such situations, historical information may be all that is available to assist the contractor's quantity surveyor, but remembering that the budget is an annual *estimate*, the more information used in the assessments, the more accurate the final judgement is likely to be.

Once the budget has been established it needs to be extended throughout the contract period. The most accurate way of doing this is to price the contract programme as accurately as possible, assessing turnover or NCP and profit returns from the operations included within each period from the known facts. If no programme is available then the contractor's quantity surveyor must use a different methodology in flexing the total figures. This can be achieved by using standard 'S'-curves or by drawing on previous experience of similar projects. The objective is to produce a total schedule that will embrace the full financial year of anticipated and secured projects.

While the contractor's quantity surveyors will usually prepare the individual project analysis, this should take place in conjunction with other team members. Contract and site managers will be able to add considerably to the assessment, as will the material-buying section.

Once these assessments have been completed they should be summarised and presented to senior management for consideration and agreement. At this juncture the budgeted company overhead will be set against the gross analysis of profit, the residue being the contractor's proposed net profit for that particular financial year.

The timing of the preparation of budget figures needs careful planning. These figures need to be prepared and agreed before the start of the financial year so that necessary actions can be appraised and any necessary action plans formulated. However, the preparation should not be so far ahead as to make the figures totally subjective. The finally agreed budget will become the tool used by the contractor to monitor performance.

The forecast summary in Figure 2.3 provides a suitable format for summarising the individual schedules. The schedule can be completed either manually or much more quickly using a simple Lotus 1-2-3 or Excel spreadsheet. As can be seen from both the individual assessments and the summary schedule, the analysis considers the NCP and the profit assessment for each project. Some contractors will differentiate between general overheads and profit. Depending on how the individual company assesses its overheads, the split may be project specific or the overhead can be dealt with as a general percentage. Whichever method is adopted, the residual figure after deduction of the overheads is the project's net assessment of profitability.

12 Commercial management in construction

Figure 2.3 Forecast summary.

The three groups to the extreme right of the summary schedule are used to analyse and compare original assessments against current forecast, with the first of the groups indicating net profit after the deduction of overhead percentages.

Forecasts

Budget figures should not be adjusted once they have been fixed. They should, however, be reviewed regularly, with the production of revised forecasts to indicate the updated situation. Most contractors will revisit their budgets on a quarterly basis and produce a forecast of the most up-to-date position, working backwards from the contractor's financial year end. This review should reanalyse all sections of the original budget, re-evaluating each project in turn, whether current, anticipated or in the final accounts stage, using the same principles as for budget preparation but with a much greater knowledge of projects in hand and how they are performing. As with the budget preparation, all elements should be analysed and reassessed, including labour, plant, material and subcontracting performance.

In addition, as projects develop the management team should develop a risk schedule or risk register for each project, locating potential problems and assessing the methodology needed to manage the risk. Various techniques can be adopted to achieve this risk analysis, but whichever technique is used, the initial development should involve the whole team in brainstorming sessions, with the resultant schedule being divided into high- and low-risk categories. This risk register is a vital tool in the management of any project and is essential to the accurate compilation of a project forecast. The risk register in Figure 2.4 may be used to schedule the results of the team sessions.

These project-specific schedules should be reviewed monthly by the team, for instance at the contract review meetings described in Chapter 7.

While the budget remains the primary mechanism for monitoring of company performance, most contractors will also monitor performance against the current forecast analysis. These forecast reviews are of vital importance to managers, since they not only are a readily demonstrable comparison with the original budget but also require the construction teams to consider each project on a regular basis, generating a greater team spirit and giving the team a further opportunity to set action plans to improve project performance.

Chapter 5 deals with the production of *cost value comparisons*. The results of these comparisons should be reviewed against the budget and forecast figures, comparing actual results of turnover and profit against the original assessments. These comparisons of actual results against budget and forecast revisions should be tabulated into three sections:

- the month currently being reviewed
- the year to date
- the total project performance to date.

Each section is analysed in turn:

- net cost of production
- gross margin
- overheads
- profit.

Murribell Builders Ltd

Risk Register

Project [   ]　　　　　　　　　　　　　Date [   ]

Risk ID Nr	Description	Likelihood	Cost Implications			Time Implications		
			Optimistic	Most Likely	Pessimistic	Optimistic	Most Likely	Pessimistic
1	Basement condition	50%	0	10000	25000	0	2 weeks	5 weeks
2	Condition of roof timbers	30%	0	2000	6000	0	4 days	2 weeks
3	Archaeologists on site	0%	0	0	2000	0	0	1 week
4	Stonework repairs	30%	500	2000	4500	0	2 weeks	5 weeks
5	Structural Issues	5%	0	0	3000	0	0	1 week
6								
7								
8								

Figure 2.4　Risk register.

Murribell Builders Ltd
Management Accounts

Date

	Actual	Month		Year to date			Total project	
	Actual	Budget	Forecast	Actual	Budget	Forecast	Actual	Budget	Forecast
Total Sales									
Tendered contracts									
Negotiated contracts									
Previous year updates									
Total									
Net cost of production									
Tendered contracts									
Negotiated contracts									
Previous year updates									
Total									
Gross profit									
Tendered contracts									
Negotiated contracts									
Previous year updates									
Total									
Overhead									
Tendered contracts									
Negotiated contracts									
Previous year updates									
Total									
Operating profit (net)									

Figure 2.5 Management account summary.

A simple table can be used to set out the comparable results (Figure 2.5). In addition to this tabular format of recording the actual results of cost value comparisons, simple graphs can be used to record the comparisons and show the current status of each project and the company as a whole. Care must be taken in compiling these graphs not to overcomplicate the end result. It is better to produce multiple graphs than attempt to put all areas under review onto one composite graph.

3 Interim Valuations

One of the main functions of the contractor's quantity surveyor is to ensure timely and full payment for work carried out on site. For most contractors interim certificates form their only source of income, from which they fund the whole of their building operation. In general, building contracts provide within their conditions regular and timely payments to the contractor.

The contract particularly referred to within this text is Joint Contracts Tribunal 1998 (JCT 1998) private edition with quantities. Within clause 30.1.1.1 of these conditions of contract there is a requirement that:

> The Architect shall from time to time as provided in clause 30 issue Interim Certificates stating the amount due to the Contractor from the Employer specifying to what the amount relates and the basis on which that amount was calculated; and the final date for payment pursuant to an Interim Certificate shall be 14 days from the date of issue of each Interim Certificate.

Clause 30.1.2 goes on further to add that:

> Interim Valuations shall be made by the Quantity Surveyor whenever the Architect considers them to be necessary for the purpose of ascertaining the amount to be stated as due in an Interim Certificate.

Timing of Interim Valuations

The architect is therefore contractually bound to issue interim certificates at the period stated within the conditions of contract.

Clause 30.1.3 details the timing requirements of interim certificates:

> Interim Certificates shall be issued at the Period of Interim Certificates specified in the Appendix up to and including the end of the period during which the Certificate of Practical Completion is issued. Thereafter Interim Certificates shall be issued as and when further amounts are ascertained as payable to the Contractor from the Employer and after the expiration of the Defects Liability Period named in the Appendix or upon the issue of the Certificate of Completion of Making Good Defects (whichever is the later) provided always that the Architect shall not be required to issue an Interim Certificate within one calendar month of having issued a previous Interim Certificate.

The contractor can therefore place a reliance upon the architect and quantity surveyor to carry out interim valuations and to issue the necessary certification as required by the contract.

What Should be Included in an Interim Valuation?

Within JCT 1998 clause 30.2 schedules the elements that should be included within each interim valuation, stating that:

The amount stated as due in an Interim Certificate, subject to any agreement between the parties as to stage payments, shall be the gross valuation as referred to in clause 30.2 less

any amount which may be deducted and retained by the Employer as provided in clause 30.4 (in the Conditions called 'the Retention') and

the amount of any advance payment or part thereof due for reimbursement stated in the Appendix pursuant to clause 30.1.1.6 and

the total amount stated as due in Interim Certificates previously issued under the Conditions.

The gross valuation shall be the total of the amounts referred to in clauses 30.2.1 and 30.2.2 less the total of the amounts referred to in clause 30.2.3 and applied up to and including a date not more than 7 days before the date of the Interim Certificate.

Clause 30.2 then sets out which amounts are the subject of retention or otherwise:

Clause 30.2.1   There shall be included the following which are subject to Retention:

the total value of the work properly executed by the Contractor including any work so executed to which Alternative B in clause 13.4.1.2 applies or to which a Price Statement or any part thereof accepted pursuant to clause 13.4.1.2 paragraph A2 or amended Price Statement or any part thereof accepted pursuant to clause 13.4.1.2 paragraph A4.2 applies but excluding any restoration, replacement or repair of loss or damage and removal and disposal of debris which in clauses 22B.3.5 and 22C.4.4.2 are treated as if they were a Variation, together with, where applicable, any adjustment of that value under clause 40. Where it is stated in the Appendix that a priced Activity Schedule relates shall be the total of the various sums which result from the application of the proportion of the work in an activity listed in the Activity Schedule properly executed to the price for that work as stated in the Activity Schedule;

the total value of the materials and goods delivered to or adjacent to the Works for incorporation therein by the Contractor but not so incorporated, provided that the value of such materials and goods shall only be included as and from such times as they are reasonably, properly and not prematurely so delivered and are adequately protected against weather and other casualties;

the total value of any materials or goods or items pre-fabricated which are 'listed items' the value of which is required pursuant to clause 30.3 to be included in the amount stated as due in the Interim Certificate;

the amounts referred to in clause 4.17.1 of Conditions NSC/C in respect of each Nominated Sub-Contractor;

the profit of the Contractor upon the total of the amounts referred to in clauses 30.2.1.4 and 30.2.2.5 less the total of the amount referred to in clause 30.2.3.2 at the rates included in the Contract Bills, or, in the case where the nomination arises from an instruction as to the expenditure of a provisional sum, at rates related thereto, or, if none, at reasonable rates.

Clause 30.2.2   There shall be included the following which are not subject to Retention:

any amounts to be included in Interim Certificates in accordance with clause 3 as a result of payments made or costs incurred by the Contractor under clauses 6.2, 8.3, 9.2, 21.2.3, 22B.2 and 22C.3;

any amounts ascertained under clause 26.1 or 34.3 or in respect of any restoration, replacement or repair of loss or damage and removal and disposal of debris which in clauses 22B.3.5 and 22C.4.4.2 are treated as if they were a Variation;

any amount to which clause 35.17 refers;

any amount payable to the Contractor under clause 38 or 39, if applicable;

the amounts referred to in clause 4.17.2 of Conditions NSC/C in respect of each Nominated Sub-Contractor.

The elements to be included within an interim certificate can therefore be summarised in less contractual terms as:

- preliminaries
- measured works
- valuation of architect's instructions or variations
- remeasurement of provisional items, prime cost (PC) sums and provisional quantities
- valuation of nominated subcontractors and suppliers
- unfixed materials on site and, where allowable, materials off site
- fluctuations, where they are allowable within the contract provisions.

Despite the obligations on the architect and quantity surveyor to produce both interim certificates and valuations, interim valuations are generally carried out together, i.e. between quantity surveyors, so that agreement can be reached between parties before certification. It is, however, far preferable to the contractor's quantity surveyor if they take the lead role themselves. Preparing the valuation in detail before meeting the client's quantity surveyor allows the contractor's quantity surveyor to model the interim valuation into the format best suited to their further requirements, i.e. subcontract accounting and cost value comparisons (see Chapter 5). In general, most quantity surveyors will go along with this methodology provided that agreement is reached at each stage in the proceedings. The benefits to the client's quantity surveyor are that many of the time-consuming tasks of valuation will be complete, such as completion of stage payment tick sheets and the counting and scheduling of unfixed materials on site, both of which could be agreed with the clerk of works should the client's quantity surveyor require.

The starting point should therefore be the establishment of interim valuation dates.

As already indicated, the contract includes, within the appendix under clause 30.1.3, the period of interim certificates, which if no other date is stipulated would be 'one month'. Therefore, the first interim valuation should be carried out one month after the 'date of possession'.

Other practical considerations should be reviewed before establishing finite dates. It may well be that the employer's organisation has set dates on which cheques are drawn. Therefore, despite the contract provisions, it may be prudent to establish interim valuation dates working backwards from the employer's payment date, provided this does not mean that the initial payment would be unduly delayed. If that were the case then it may be necessary, with the employer's assent, to agree an initial valuation earlier than previously described or allowed within the conditions of contract.

Given the date for possession, the first valuation may fall when either the client's quantity surveyor or contractor's quantity surveyor has commitments on other projects. It may again, therefore, be prudent to delay the initial valuation so that events as indicated above can be overcome.

Once the initial date has been established it is easy enough to schedule future interim valuation dates and confirm these dates in writing to the client's quantity surveyor and architect. At this juncture the contractor's quantity surveyor should advise all subcontractors of the dates and also formally advise the contractor's costing department in order that cost information can be provided as expeditiously as possible, remembering to advise those subcontractors of the dates when their orders are placed later in the project.

An important consideration in establishing valuation dates is that of cash flow. Cash flow is absolutely vital to any business and the monitoring of cash balances within a contracting organisation is paramount to the company's success, particularly given that, in the main, contractors are paid substantially in arrears of works being completed on site.

The understanding of the payment system is very important for the contractor's quantity surveyor, not only in terms of payment for contract works once the project is up and running, but also in terms of what is becoming more prevalent, that is the requirement of precontract payments. Such payments relate to payments for statutory authority fees, planning and building regulation fees or consultant's fees if they form part of the contract. Many of the above types of payment are required either very early in the contract administration or in some cases considerably in advance of contract commencement to enable an effective start on site. For example, the initial contract operations may revolve around main service diversions. If that is the case, then either prepaid orders will have to be placed by the client directly or special provision will have to be made within the conditions of contract to allow for such a situation if the employer wishes these sums to be paid by the contractor.

In addition to the above, consideration should be given to the requirements of the contractor's accounting system. To ensure good cash flow within any contracting organisation it is important to consider income and expenditure requirements. For example, salaries and associated costs are generally by far the largest part of most contractors' overheads and will be paid out on a set date every month. It is important therefore that such matters are considered when establishing dates for interim valuations.

Similarly, credit arrangements will be in place with suppliers which generally require that payment is made 30 days from the end of the month when delivery takes place. Looking at this scenario, therefore, consider that most contracts allow for payment 14 days after the 'date of issue' of the architect's interim certificate and that in the case of JCT 1998 the interim certificate must, as required by clause 30.2, include the value of works as described previously up to a period not more than 7 days before the date of the interim certificate. It then follows that if the contractor arranges interim valuation dates around the above criteria and agrees dates at the end of the first week of any month, then payment will be forthcoming from the employer before that month end, helping the contractor's cash flow in that suppliers will not require payment until the end of the following month.

Method of Production

Given that agreement can be reached with the client's quantity surveyor allowing the contractor's quantity surveyor to take the lead in the initial preparation of the interim valuation, it will be necessary during the lead-in time to the project and before the initial interim valuation becomes due, for the contractor's quantity surveyor to establish how he proposes to value the project and agree that methodology with the client's quantity surveyor.

Again, there are many considerations that are very much dependent on the type of project being constructed. If it is a housing project, whether new or refurbishment, the valuation of the project will probably lend itself to the use of stage payments (a system of analysing the bills of quantity into various work groups). If it is a one-off project such as a garage, school or a factory, where there is no repetition, then it will be necessary to value the project by scrolling the bills of quantity item by item to record elements of completed work, measuring work sections as the works proceed.

Looking first at the stage payment method of valuation, it is necessary to understand how the project is to be constructed before establishing the stage descriptions, and detailed discussion with the project manager will be most valuable at this juncture. Once these discussions have taken place then a schedule can be produced as near as possible to the build programme order.

It is of paramount importance at this point that accurate stage values are made. These initial calculations will form the basis of all future interim payments which, in turn, form the starting point of the contractor's internal costing process. All elements of each work stage need to be established, located within the bills of quantities and included within the stage value. It is important at this stage to reconcile the stage payment totals now established with the original bill of quantity totals to ensure that all areas have been allocated.

A point worth noting is that the more stages there are and the smaller the stage value, the more accurate the assessment will be when the valuation is complete. The downside to this is that on-site work in completing the valuation will take longer. Therefore, a balance needs to be drawn at this juncture.

Murribell Builders Ltd

Analysis of Stage Payments

Project: _____

Bill of Quantity Reference   **Description**

Page	Item	Brickwork 1st lift	Windows Ground Floor	Brickwork 2nd lift	Floor Joists	Brickwork 3rd lift	Windows First Floor	Brickwork 4th lift	Roof Trusses	Brickwork Peaks
1/9	1-9	1025.76		1025.76		1025.76		1025.76		1025.76
1/10	1-6	279.37		279.37		279.37		279.37		279.37
	7-12		93.62				93.62			
2/1	1-5		2150.35				2150.35			
2/2	1-8				897.99					
	9-14								998.36	

Figure 3.1 Stage payment schedule.

Once the stage descriptions have been established it is necessary to abstract from the bill of quantities the various billed items and values required for each stage (Figure 3.1). A thorough construction knowledge of the project is required at this point so that all the requisite elements of the bills of quantities can be included in the appropriate stage value. For example, the contractor's quantity surveyor needs to remember that lintels go into the second and fourth lift brickwork stages and that the wallplate goes with the fourth lift brickwork stage, including the necessary holding down straps, and so on, with an allowance left for work to peaks, a brickwork operation generally left until after roof trusses have been installed.

The need for great accuracy in the initial compilation of the stage contents and values cannot be overstated. If, for example, the units being constructed include the provision of an entrance porch of similar construction to the remainder of the unit, scaffolding restrictions during the construction of the main unit would restrict porch construction. The porch would therefore be constructed at a later date once roofworks were complete and original scaffolding had been removed or adapted. This being the case, the stages must account for this situation in terms of both value and timing of the porch. It is also worthwhile checking the project method statements to check the proposed construction methodology.

As can be seen from the stage payment schedule in Figure 3.2, the units or addresses are listed along the top. The valuation of the units then becomes a

Murribell Builders Ltd

| Project | | | | | | | | | | | | Date | | |
| Interim Valuation Nr. | | | | | | | | | | | | In attendance | | |

Stage	1	2	3	4	5	6	7	8	9	10	11	12	units complete	stage value	total
Brickwork 1st lift															
Windows Ground floor															
Brickwork 2nd lift															
Floor joists															
Brickwork 3rd lift															
Windows First floor															
Brickwork 4th lift															
Roof trusses															
Brickwork peaks															
Roof Tiling															
Joiner 1st fix															
Plumber 1st fix															
Electrician 1st fix															
Plasterwork															
Kitchen fittings															
Joiner 2nd fix															
Plumber 2nd fix															
Electrician 2nd fix															
Painterwork															
Wall Tiling															
Floor Tiling															
Final Clean															
Handover															
														Total	

Figure 3.2 Stage payment analysis (part example).

simple exercise of a site visit to each unit to analyse work in progress, each unit section being reviewed and entered as a tick if the unit is complete or as a percentage. Then a simple calculation is used to assess the total value completed at any given time for inclusion within the interim valuation. With today's computer facilities, as in many areas of scheduling work, a very basic spreadsheet system can be used to produce this information, making valuations using this system easy to complete.

As an enhancement of this system, if each interim valuation is colour coded the documentation provides a permanent record of when each element was carried out during the contract period.

While the example shown in Figure 3.1 only reflects the stage payments for an actual superstructure element of measured work of a unit, there is no reason why all elements of the project cannot be incorporated into the stage payment schedule, from preliminary items through to drainage and site works, provided that there is a uniform build system throughout the project. Non-standard items will require remeasurement as works proceed or they can be included by simply scrolling the bill of quantities, scheduling out completed items.

This brings us to the second method of valuing measured work sections, which is working or scrolling through the project item by item and assessing quantities or percentage completion against each item or group of items. While this may seem a daunting task, provided that site valuation notes are full and accurate and remeasurement schedules are up to date before starting the valuation, the task should not prove too difficult to complete.

Using this technique many quantity surveyors, both contractor's and client's, will mark up the bill of quantities on site as items are considered and booked off as complete, or in the case of partially completed items an amendment to the quantity will be made for valuation purposes. The interim valuation can then be written up or inputted onto a computer at a later date.

The methodology of assessment will vary depending on the type of project and item under consideration. Items such as doors and windows can simply be enumerated and if the contractor's quantity surveyor keeps his window and door schedule up to date then the valuation of these types of item can be kept fairly simple.

Large areas such as brick and blockwork, however, will require site measuring as works proceed if an accurate valuation is to be made. As with most of a quantity surveyor's tasks, detailed records will make the job a lot simpler. Marking up drawings on site to show areas of brickwork completed will always prove a valuable document in producing and agreeing the valuation. Once a quantity has been established then this can be used for the valuation assessment, either as an accurate quantity in its own right or reduced to a proportion or percentage of the whole brickwork item. This assessment can thereafter be used and linked to similar items for valuation purposes. Care needs to be exercised when using percentages with these assessments, as external brickwork, which may include expensive bricks, cills and lintels, may be completed at a slower pace than the internal skin of blockwork or the internal blockwork itself, rendering an overall percentage ineffective.

Drainage and service installation, for example, also benefit from keeping a marked-up drawing. For measurement purposes, a drainage schedule should be used either for taking off the drainage completed at the time of each interim valuation, if the scheme is installed as per the drawings, or for recording actual works completed on site if the drainage works are measured provisionally within the original bills of quantities.

When carrying out the interim valuation it is therefore a case of working through the bill of quantities, item by item, and making the required assessment; for example:

Page 1–14		Substructures	All complete	47,613.67
Page 15	Item 1	Brickwork	51 m^2 £62.72	3,198.72
Page 16	Item 1–9	Blockwork	27%	12,654.56
Page 17	Item 1–16	Sundry brickwork	27%	851.29
Page 31	Item 4	Windows	6 nr. £432.45	2,594.70

There is no need to use one single method of assessment; rather, each item should be judged on its own merit. To aid further work on the document and to facilitate checking by the client's quantity surveyor, the more detail provided, the easier subsequent tasks will become. There are many ways in which this assessment can be carried out. Some quantity surveyors will mark the bills of quantity on site with the valuation assessments as each item is considered, writing up the valuation for presentation back at the office. Others will produce pro-formas of the bills of quantities which will be updated with each valuation, while yet others will simply work through the bills of quantities and write up the valuation on site. All methods are acceptable provided care is taken with accuracy and the methodology used in the compilation of any interim valuation considers any further uses for which the valuation analysis will be required (see Appendix 1).

Several computerised valuation packages are available to the contractor's quantity surveyor. One such system is the Conquest valuation package. Conquest, a specialist company dedicated to the construction industry, has been developing this system for about 20 years. Conquest continuously develops its construction packages by constant interaction between the users of the software, enabling the systems to be fine-tuned to suit industry requirements. With these systems there is no reinvention of the wheel. They are based around standard industry working practices as detailed within this chapter and have been developed by experienced estimators and quantity surveyors. The current edition, Conquest for Windows, was launched in 1998 and updated in 1999. It uses simple-to-operate Windows technology to provide a very user-friendly package.

While each Conquest system is stand-alone, the valuation system is better linked to the estimating and quotation comparison packages. Original information can be imported by several methods. Bills of quantities can be scanned in using optical character recognition (OCR) or on a floppy disk or CD, but more information will be available to the contractor's team if the valuation system is generated from the estimating package.

The main benefits of any computerised valuation systems is that of speed of production and the ability of the systems to generate a myriad of back-up information for use not only by the contractor's quantity surveyor but also by the whole management team. Valuations can produce four separate quantities against each item:

- the quantity from the original bill
- the quantity from last month's valuation

- the quantity entered this month
- the difference between this month's quantity and last month's quantity.

This is a useful tool for the contractor when analysing output on any projects.

When linked to the estimating package the valuation system gives the added benefit to the contractor's team that the facility is available to analyse the original tender without the need for considerable calculation. Resources can be extracted such as:

- list all items using a JCB or a Hymac
- list all items with a value of more than £x
- list all items with a quantity over 50
- list all items belonging to Excavation or Drainage
- list all items of Brickwork with a value greater than £x.

These figures can then be used by the contractor's team as part of the management process. Similarly, once valuations have been completed similar information can be extracted to compare against original allowances and current costing. Labour, plant, material and subcontract figures can all be readily extracted to compare against actual job cost. The system produces actual figures as well as pie or bar charts showing the value of each resource as a lump sum or percentage of the overall project.

The production of valuations is no different to that described elsewhere in this chapter. The systems are very flexible: elements can be grouped together or taken singly, and can be expressed as percentages or quantity; for example:

- Bill 1 Page 3: 25%
- Bill 2 Pages 3 to 4: 30%
- Items 3/1/A to 3/2/F inclusive: 17%
- Items 4/2/B,C,F,H,K: 75%
- Items belonging to the activity 'First Lift Brickwork': 50%.

The system allows for items to be added in an addendum bill or at any point within the original document and also includes within the software package the provision for 'taking off' during the valuation process. As with the Conquest estimating system, taking off can be done manually or via a digitiser.

Preliminaries

Several formats can be used for the valuation of the preliminary section of a project. Preliminaries are generally site-specific matters that are globally required for each project, such as site management, accommodation and welfare, scaffolding, general plant and security, i.e. those matters that cannot readily be priced within each specific item in the bills of quantities.

For the purposes of the interim valuation, preliminaries evaluation will depend on the level of detail in the pricing of the original document and what the contractor's estimator has included in their assessment. Some contractors will include preliminaries as a lump sum which makes accurate assessment impossible; some will include all the project profit; whereas others will spread either all or part of their profit throughout the measured section of the bill of quantities.

Interim valuations 27

Murribell Builders Ltd

| Project | City Centre Hotel | | Date | |
| Interim Valuation Nr. | 1 | | In Attendance | |

Preliminaries

Total amount included within Bills of Quantities
              page   1/48   £148,375.00

Deduct
Prime cost sums	1/34-37	-41000.00	
Provisional sums	1/39	-8200.00	
Dayworks	1/40	-5000.00	
		-54200.00	148375.00
			-54200.00
Net Amount of Preliminaries			**£94,175.00**

Preliminary Analysis

	Set up	Duration	Completion	Valuation
	a	b	c	
Project start up	2000			2000.00
Site accommodation	1200			1200.00
Temporary roads	2100		400.00	1700.00
Site service				
Water	850			850.00
Electric	1200			1200.00
Telephone	320			320.00
Scaffolding		24800		
Final clean			750.00	
Clear site			1800.00	
	7670	24800	2950.00	7270.00
	total a+b+c		35420.00	

Residue of preliminaries to be spread over full contract period of 40 weeks

Net amount of Preliminaries	94175.00			
Deduct				
Total a+b+c	35420.00			
	58755.00			

| Valuation as at week nr | 4 | 4/40 | 58755.00 | 5875.50 |

| Total Preliminaries for Valuation Nr. | 1 | | | £13,145.50 |

Figure 3.3 External preliminary draw sheet.

There is no reason why preliminaries cannot be included within a schedule of stage payments; however, more accurate assessment is preferable, as explained below. It is therefore customary to treat the valuation of preliminaries separately at the outset of each interim valuation (Figure 3.3).

Individual items are claimed as a percentage of their total value or, as is more generally the case, the preliminaries are subdivided into their 'fixed' or 'time related' sections and the amounts apportioned to each interim valuation.

This method allows point load items, such as site establishment costs and scaffolding, to be fully included within the interim valuation, relative to when they are expended. It allows for the retention of monies for site clearance and final cleaning, and facilitates the time-related items to be spread over the whole of the contract period. The methodology satisfies the needs of both quantity surveyors. The employers will be reassured that their quantity surveyor is not paying in advance and the contractor's quantity surveyor can be happy in that he will be paid for preliminary items when they are required and, importantly, when they need to be paid for by the contractor.

Larger items, such as scaffolding, cranes, site establishment and supervision costs, will require careful assessment by both parties. As already indicated, scaffolding should be included as and when it is used, which on a complex site could prove to be a very interesting assessment; craneage, although similar, may be a little easier to identify. Site establishment costs on larger schemes could also be complicated. Does the site need extensive new temporary supplies? What IT requirement have been allowed for? Has the accommodation been specially purchased for the scheme rather than hired? Has there been a need to purchase any major elements of plant specifically for the project? With regard to supervision, has the contractor's team been working in advance of contract commencement and do they now require recompense from the preliminaries allowance? All such matters will need to be addressed during the initial debates relating to preliminaries allocation.

If a lump sum only is included within the bills of quantities, valuation can take one of two forms. The total can be assessed on a programme basis, i.e. an amount per week by simply dividing the total preliminaries amount by the contract period, or the amount can be claimed in the same proportion as the interim valuation would be to the original contract sum. While this methodology is widely used, it is far from satisfactory to both the client's and the contractor's quantity surveyors. The respective quantity surveyors may agree an arbitrary split of such lump sums, but only the contractor will know the exact breakdown of the original figure, meaning that the client's quantity surveyor may take a cautious view when agreeing and assessing figures. While there is no real need for 100% accuracy in assessing the preliminaries figure, the quantity surveyor has a duty to his client to ensure that overpayments are not made to the contractor at this stage, or indeed at any stage during the contract.

However, to obtain accurate cost value comparisons it is also necessary for the contractor's quantity surveyor to produce a finite schedule of preliminary costings. The internal preliminary draw schedule shown in Figure 3.4 is sufficient for most projects but can be easily adjusted to include other more project-specific items if necessary. This schedule also indicates the comparison between internally and externally claimed preliminaries. The calculated difference is transferred to the valuation adjustment schedule discussed in Chapter 5.

Variations

The conditions of contract, under clause 4, Architect's Instructions, allow for the architect to vary the works should the need arise. The conditions of contract also advise under clause 4.3.1 that all instructions must be issued *in writing*. While there are other provisions within the contract documentation that allow for payment of work without an official architect's instruction, to avoid doubt the contractor will be better placed when payment is sought through the interim valuation process if an architect's instruction has been issued. Failure to obtain an architect's instruction therefore may debar the contractor from receiving rightful payment for works carried out on site.

Murribell Builders Ltd

Internal Preliminary Schedule

Project [   ]   Date [   ]

	Total Project Preliminaries			Valuation Nr.	
	Weeks	Amount per week	Total	Week Nr.	Total
Site Manager			0.00		0.00
General Foreman			0.00		0.00
Trades Foreman			0.00		0.00
Trades Foreman			0.00		0.00
Security			0.00		0.00
Temporary Roads			0.00		0.00
Site Offices			0.00		0.00
Transportation			0.00		0.00
Furniture			0.00		0.00
Mess Cabins			0.00		0.00
Transportation			0.00		0.00
Furniture			0.00		0.00
Tool Vault			0.00		0.00
Toilets			0.00		0.00
Stores			0.00		0.00
Compound Fencing			0.00		0.00
Compound Gates			0.00		0.00
Telephone Rental			0.00		0.00
Telephone Calls			0.00		0.00
Temporary Water			0.00		0.00
Temporary Electric			0.00		0.00
Electricity Running Charges			0.00		0.00
Heating			0.00		0.00
Sign Boards			0.00		0.00
Cranes Fixed			0.00		0.00
Cranes Mobile			0.00		0.00
Mixers			0.00		0.00
Mixer Set Up			0.00		0.00
Pumps			0.00		0.00
Compressors			0.00		0.00
Roller			0.00		0.00
Wacker Plate			0.00		0.00
Kangos			0.00		0.00
Drills			0.00		0.00
Scaffolding External Walls			0.00		0.00
Internal Walls			0.00		0.00
Stairwells			0.00		0.00
Chimney			0.00		0.00
Clear Site			0.00		0.00
Final Clean			0.00		0.00
Drying Out			0.00		0.00
Protect Roadways			0.00		0.00
Winter Working			0.00		0.00
Overtime			0.00		0.00
Service Gangs			0.00		0.00
Concrete Testing			0.00		0.00
Protection			0.00		0.00
Small Tool Allowance			0.00		0.00
Total Internal Preliminaries			**£0.00**		**£0.00**
Total External Preliminaries			**£0.00**		**£0.00**
Balance to Valuation Adjustment Schedule					**£0.00**

Figure 3.4 Internal preliminary draw sheet.

Clause 4.1.1 requires that the contractor must forthwith comply with all instructions issued by the architect, and that should the contractor not comply with these instructions the architect is empowered under clause 4.1.2 to instruct others to carry out the instruction should the contractor's default continue for a period of more than 7 days. Clause 4.1.2 also allows for adjustment of the contractor's account in the event of such default. Clause 4 does, however, include provisions for the contractor to query the issue of certain instructions.

In most contractors' organisations it is part of the contractor's quantity surveyor's duties to seek out any variations and to ensure that an architect's instruction is issued accordingly, within the time constraint set out within clause 4.

Clause 13, *Variations and provisional sums*, defines the term variation as:

> the alteration or modification of the design, quality or quantity of the Works including:
>
> > the addition, omission or substitution of any work,
> >
> > the alteration of the kind or standard of any of the materials or goods to be used in the Works,
> >
> > the removal from the site of any work executed or materials or goods brought thereon by the contractor for the purposes of the Works other than work materials or goods not in accordance with the contract;
>
> the imposition by the Employer of any obligation or restrictions in regard to the matters set out in clauses 13.1.2.1 to 13.1.2.4 or the addition to or alteration or omission of any such obligation or restrictions so imposed or imposed by the Employer in the Contract bills in regard to:
>
> > access to the site or use of any specific parts of the site,
> >
> > limitations of working space,
> >
> > limitation of working hours,
> >
> > the execution or completion of the work in any specified order.

As indicated at the start of this chapter, it is a prime function of the contractor's quantity surveyor to ensure timely and *full* payment for works carried out by the contractor. They must therefore understand fully the requirements of the above extract from clause 13.

In addition, the contractor's quantity surveyor must understand fully the provisions contained in clause 13.5, which sets out the methods that are to be used for the valuation of variations. These are:

> where the additional or substituted work is of similar character to, is executed under similar conditions as, and does not significantly change the quantity of, work set out in the Contract Bills the rates and prices for the work so set out shall determine the Valuation;
>
> where the additional or substituted work is of similar character to work set out in the Contract Bills but is not executed under similar conditions thereto and/or significantly changes the quantity thereof, the rates and

prices for the work so set out shall be the basis for determining the valuation and the valuation shall include a fair allowance for such difference in condition and/or quantity;

where the additional or substituted work is not of similar character to work set out in the Contract Bills the work shall be valued at fair rates and prices;

where the Approximate Quantity is a reasonably accurate forecast of the quantity of work required the rate or price for the Approximate Quantity shall determine the Valuation;

where the Approximate Quantity is not a reasonably accurate forecast of the quantity of work required the rate or price for that approximate Quantity shall be the basis for determining the Valuation and the Valuation shall include a fair allowance for such difference in quantity.

The last two conditions apply to the extent that the work has not been altered or modified other than in quantity.

To the extent that the Valuation relates to the omission of work set out in the Contract Bills the rates and prices for such work therein set out shall determine the valuation of the work omitted.

Clause 13.5.3 adds further clarification to those variations valued under conditions 13.5.1 and 13.5.2:

measurement shall be in accordance with the same principles as those governing the preparation of the Contract Bills;

allowance shall be made for any lump sum adjustments in the Contract Bills; and

allowance, where appropriate, shall be made for any addition to or reduction of preliminary items of the type referred to in the Standard Method of Measurement, 7th Edition, Section A (Preliminaries/General Conditions); provided that no such allowance shall be made in respect of compliance with an Architect's Instruction for the expenditure of a provisional sum for defined work.

Clause 13.5.4 deals with the valuation of additional or substituted work which cannot be properly valued by the measurement rules scheduled above:

the prime cost of such work (calculated in accordance with the 'Definition of Prime Cost of Daywork carried out under a Building Contract' issued by the Royal Institution of Chartered Surveyors and the Building Employers Confederation (now Construction Confederation) which was current at the Base Date) together with percentage additions to each section of the prime cost at the rates set out by the Contractor in the Contract Bills; or

where the work is within the province of any specialist trade and the said Institution and the appropriate body representing the employers in that trade have agreed and issued a definition of prime cost of daywork, the prime cost of such work calculated in accordance with that definition which was current at the Base Date together with percentage additions on the prime cost at the rates set out by the Contractor in the Contract Bills.

Provided that in any case vouchers specifying the daily time spent upon the work, the workmen's names, the plant and the materials employed shall be delivered for verification to the Architect or his authorised representative not later than the end of the week following that in which the work has been executed.

As the conditions of contract describe, the rates and prices set out in the contract bills can be used in two situations, i.e. where the work being valued relates to omissions, and in the case of additions, only where the work being carried out is of similar character, carried out under similar conditions and where the quantity is not unduly varied. If, for example, the quantity of doors is marginally increased, but using the project-specified doors, then it would be appropriate to use existing contract bill rates, provided the doors are similarly positioned throughout the project. Similarly, if quantities of foundation concrete increased marginally over the contract bill allowance then again the rates and prices within the contract bills would prevail.

If, however, the three conditions noted above are not met, then the contractor is entitled to a revaluation of the work being undertaken. Using the door scenario again, if the size or thickness of the doors increases then the existing rate can be used, with an adjustment being made to the material element of the contract rate to account for the differing size. If, however, the specification changes from paint quality to hardwood self-finished doors, then other elements of the existing rate may require adjustment. Clearly, the material content will need adjustment and in this example the labour content will need considerable adjustment. The hanging operation of the door will require more time and care, and further adjustment will be required to account for the extra care needed in storing the doors and in moving them to their position for final fixing on site. Changing the specification to self-finish doors may also mean that deliveries are affected and delays to completion may ensue. There is no reason why the effect of these delays cannot be accounted for within the valuation of this variation, extending the date for completion as required as a separate matter.

Most revisions in specification and changes in situation or quantity can be valued using the above methodology, but there will be instances when adjusting the existing rate is not practical and a fair valuation should be agreed between quantity surveyors.

A fair valuation does not mean that the contractor's quantity surveyor has *carte blanche* to cover all costs relating to the valuation of such variations. Quantity surveyors representing the employer will contend that there should still be a regard for the original pricing level. They may look, for example, at using existing built-up rates for labour, even though the labour output constant for the operation may differ significantly.

It is, however, general practice that the client's quantity surveyor will not take advantage of pricing errors within the contract bills. If the contractor has incorrectly priced any item of work then, while they are obliged to carry out such works up to the level of quantum contained in the contract bills, any additional quantity carried out should be revalued at a fair rate for the job.

To aid agreement, the contractor's quantity surveyor should have all information to hand when attempting to agree rates with their opposite number. Details such as time allocation sheets, material invoices or delivery records and detailed plant returns will assist greatly in the agreement of the valuation of such variations.

The next element to be considered is that of work to be valued using its *'prime cost'*. This method is only applicable when the valuation of works being carried out cannot be valued using one of the previously described methods. This system uses, as a method of calculation, the rules set out in the 'Definition of Prime Cost of Daywork carried out under a Building Contract'.

Valuation on *'daywork'* requires accurate records of labour, plant and material to be kept on a regular basis. Although it is not a contract provision many contracts are amended to include a condition to the effect that the contractor must notify the architect or his appointed representative before starting any work that is to be valued on a daywork basis. The contract does, however, require that the daywork vouchers specifying the daily times spent on the work are delivered for verification to the architect or his representative (generally the clerk of works) not later than the end of the week following that in which the work was carried out. These records should also include the workmen's names, together with the plant and materials used during the operation. Although it is not a contractual requirement, it is advisable to include each operative's trade alongside their name, since not only will this make the architect's job of verification easier but it will also assist greatly in the contractor's quantity surveying in valuing the daywork voucher at a later date ready for inclusion within the variation account.

The recording of works to be valued on a daywork basis is best carried out in conjunction with the site manager or general foreman, using their specific knowledge of each operation and using the operatives' daily allocation sheets, material and plant returns as a further source of recorded information to assist in completing accurate records. In many instances the site manager or general foreman will already have marked up the allocation sheets, etc., ready for discussion with the quantity surveyor.

Valuation of daywork vouchers must be completed strictly in accordance with the rules set out in the 'Definition of Prime Cost of Daywork'. This document sets out the elements of what should be included in the rates for labour, material and plant sections of the daywork voucher, and also what can be included in relation to an addition for incidental costs, overheads and profit. A typical example of a daywork voucher is shown in Figure 3.5.

The definition sets out the rules as follows.

Section 2

Composition of total charges

    2.1 The prime cost of daywork comprises the sum of the following costs:

        (a) Labour as defined in Section 3.

        (b) Material and Goods as defined in Section 4.

        (c) Plant as defined in Section 5.

Daywork Voucher

MURRIBELL BUILDERS LTD

Figure 3.5 Daywork voucher.

    2.2 Incidental costs, overheads and profit as defined in Section 6, as provided in the building contract and expressed therein as a percentage adjustments applicable to each of 2.1 (a)–(c).

Section 3

Labour

    3.1 The standard wage rates, emoluments and expenses referred to below and the standard working hours referred to in 3.2 are those laid down for the time being in the rules or decisions of the National Joint Council for the Building Industry and the terms of the Building and Civil Engineering Annual and Public Holiday agreements applicable to the works, or the rules or decisions or agreements of such body, other than the National Joint Council for the Building Industry, as may be applicable, relating to the class of labour concerned at the time when and in the area where the daywork is executed.

3.2 Hourly base rates for labour are computed by dividing the annual prime cost of labour, based upon standard working hours and as defined in 3.4 (a)–(i), by the number of working hours per annum (see examples below).

3.3 The hourly rates computed in accordance with 3.2 shall be applied in respect of the time spent by operatives directly engaged on the daywork, including those operating mechanical plant and transport and erecting and dismantling other plant (unless otherwise expressly provided in the building contract).

3.4 The annual prime cost of labour comprises the following:

(a) Guaranteed minimum weekly earnings (e.g. Standard Basic Rate of Wages, Joint Board Supplement and Guaranteed Minimum Bonus in the case of NJCBI rules).

(b) All other guaranteed minimum payments (unless included in Section 6).

(c) Differentials or extra payments in respect of skill, responsibility, discomfort, inconvenience or risk (excluding those in respect of supervisory responsibility – see 3.5)

(d) Payment in respect of public holidays.

(e) Any amounts which may become payable by the Contractor to or in respect of operatives arising from the operation of rules referred to in 3.1 which are not provided for in 3.4 (a)–(d) or in Section 6.

(f) Employer's National Insurance contributions applicable to 3.4 (a)–(e).

(g) Employer's contribution to annual holiday credits.

(h) Employer's contributions to death benefits scheme.

(i) Any contribution, levy or tax imposed by statute, payable by the contractor in his capacity as an employer.

3.5 Note

Differentials or extra payments in respect of supervisory responsibility are excluded from the annual prime cost (see Section 6). The time of principals, foremen, gangers, leading hands and similar categories, when working manually, is admissible under this section at the appropriate rates for the trades concerned.

Section 4

Materials and goods

4.1 The prime cost of material and goods obtained from stockists or manufacturers is the invoice cost after deduction of all trade discounts but including cash discounts not exceeding 5 per cent and includes the cost of delivery to site.

4.2 The prime cost of materials and goods supplied from the Contractor's stock is based upon the current market prices plus any appropriate handling charges.

4.3 Any Value Added Tax which is treated, or is capable of being treated, as input tax (as defined in the Finance Act, 1972) by the Contractor is excluded.

Section 5

Plant

5.1 The rates for plant shall be as provided in the building contract.

5.2 The costs included in this section comprise the following:

(a) Use of mechanical plant and transport for the time employed on daywork.

(b) Use of non-mechanical plant (excluding non-mechanical hand tools) for the time employed on daywork

5.3 Note: The use of non-mechanical hand tools and of erected scaffolding, staging, trestles or the like is excluded (see Section 6).

Section 6

Incidental costs, overheads and profit

6.1 The percentage adjustments provided in the building contract, which are applicable to each of the totals of Sections 3, 4 and 5, comprise the following:

(a) Head Office charges.

(b) Site staff including site supervision.

(c) The additional cost of overtime (other than that referred to in 6.2).

(d) Time lost due to inclement weather.

(e) The additional cost of bonuses and all other incentive payments in excess of any guaranteed minimum included in 3.4 (a).

(f) Apprentices' study time.

(g) Subsistence and periodic allowances.

(h) Fares and travelling allowances.

(i) Sick pay or insurance in respect thereof.

(j) Third party and employer's liability insurance.

(k) Liability in respect of redundancy payments to employees.

(l) Employer's National Insurance contributions not included in Section 3.4.

(m) Tool allowances.

(n) Use, repair and sharpening of non-mechanical hand tools.

(o) Use of erected scaffolding, staging, trestles or the like.

(p) Use of tarpaulins, protective clothing, artificial lighting, safety and welfare facilities, storage and the like that may be available on the site.

(q) Any variation to basic rates required by the Contractor in cases where the building contract provides for the use of a specified schedule of basic plant charges (to the extent that no other provision is made for such variation).

(r) All other liabilities and obligations whatsoever not specifically referred to in this section nor chargeable under any other section.

(s) Profit.

6.2 Note: The additional cost of overtime, where specifically ordered by the Architect/Supervising Officer, shall only be chargeable in the terms of prior written agreement between the parties to the building contract.

Build-up of Standard Hourly Base Rates – Applicable at 25 June 2001

Under the JCT Standard Form, dayworks are calculated in accordance with the Definition of the Prime Cost of Dayworks carried out under a Building Contract published by the RICS and the NFBTE. The following build-up has been calculated from information provided by the Building Cost Information Service (BCIS) in liaison with the NFBTE. The example is for the calculation of the standard hourly base rate, is for convenience only and does not form part of the definition; all basic rates are subject to re-examination according to when and where the dayworks are executed.

Standard working hours per annum

52 weeks at 39 hours		2028	hours
Less: 21 days annual holiday			
16 days at 8 hours	128		
5 days at 7 hours	35		
8 days public holiday			
7 days at 8 hours	56		
1 day at 7 hours	7	226	hours
		1802	hours

38 Commercial management in construction

Guaranteed minimum weekly earnings			Craftsman £		Labourer £
Standard basis rate			261.30		196.56
Guaranteed minimum earnings			–		–
Joint Board supplement			–		–
			_261.30_		_196.56_

Hourly Base Rate	Rate £	Craftsman (£6.70)	Rate £	General Operative (£5.04)
Guaranteed minimum weekly earnings: 46.21 weeks at	261.30	12074.67	196.56	9083.04
Extra payment for skill, responsibility discomfort, inconvenience or risk*: 1802 hours	–	–	–	–
Employer's National Insurance Contribution**: 11.9% of	8053.94	958.42	5062.31	602.42
Holidays with pay: 226 hours at minimum rate	6.70	1514.20	5.04	1139.04
***CITB Annual Levy: 0.5% of	13588.87	67.94	10222.08	51.11
Welfare benefit: 52 stamps at £2.20		114.40		114.40
Public holidays (included with Holiday with Pay above)		–		–
		£14729.63		£10990.01

Hourly Base Rate

Craftsman
$$\frac{£14729.63}{1802} = \mathbf{£8.17}$$

General Operative
$$\frac{£10990.01}{1802} = \mathbf{£6.10}$$

Notes

*Only include in hourly base rate for operatives receiving such payments.

**National Insurance Contribution based on 11.9% above the Earnings Threshold for 46.21 weeks at £87.01 (= £4020.73).

***From 19 March 2001, the CITB levy is calculated at 0.5% of the PAYE payroll of each employee plus 1.50% of payments made under a labour-only agreement. Building contractors having a combined payroll of £61000 or less are exempt. The levy is included in the daywork rate of each working operative, the remainder being a constituent part of the overheads percentage.

****From 29 June 1992, the Guaranteed Minimum Bonus ceased and is consolidated in the Guaranteed Minimum Earnings.

Holidays: The winter holiday is 2 calendar weeks taken in conjunction with Christmas, Boxing and New Year's Day.

The Easter holiday shall be four working days immediately following Easter Monday.

The summer holiday shall be two calendar weeks.

The possible additional one day holiday for June 2002 is not included in the above calculation, where this is considered applicable a further 3p for a Craftsman and 2p for General Operatives should be added to the above rate.

Daywork, Overheads and Profit

Fixed percentage additions to cover overheads and profit no longer form part of the published daywork schedules issued by the Royal Institution of Chartered Surveyors, and contractors are usually asked to state the percentages they require as part of their tender. These will vary from firm to firm and, before adding a percentage, the list of items included in Section 6 of the above schedule should be studied and the cost of each item assessed and the percentage overall addition worked out.

As a guide, the following percentages were extracted from recent tenders for contract work:

- on labour costs 80–200%
- on material costs 10–20%
- on plant costs 15–30%.

Much higher percentages may occur on small projects or where the amount of daywork envisaged is low.

A separate file should be kept by the contractor's quantity surveyor of all such architect's instructions issued and a fully updated schedule should be maintained at the front of the architect's instruction file. Behind each separate instruction the contractor's quantity surveyor should hold a record of all necessary remeasurements relating to the particular variation, cross-referenced to the dimensions book if applicable, and also any further information such as the contractor's invoices, time sheets or plant records, or subcontract accounts used in the costing of the variation.

The costing and agreement of the valuation of architect's instructions with the client's quantity surveyor must be carried out regularly and the summary schedule updated before each interim valuation. As variations are agreed with the client's quantity surveyor the schedule should be marked or highlighted accordingly, thus eliminating the need to revisit agreed areas of the project account.

The rules and conditions relating to how and when architect's instructions can be issued were set out at the beginning of this section, but who instigates the variation?

As most projects develop the architect will issue further drawings which may be required to amplify the detail, revise the construction method or amend the specification to be used. Such amendments will generally be issued with an architect's instruction which will detail all of the revisions or alterations that have occurred.

Many architect's instructions will be issued following requests made by the contractor either as confirmation of site instructions issued by the clerk of works under clause 12 or as confirmation of the contractor's written requests for same. The contractor's quantity surveyor can play a large part in the generation of such requests, remembering that one of his main aims is to ensure full payment for works carried out. The contractor's quantity surveyor should check all areas of each project to establish whether variations have occurred and where an architect's instruction is required. There are many ways in which this can be achieved.

As drawings are issued the contractor's quantity surveyor should check these against original tender drawings, looking for any amendments that may have occurred. As already indicated, most architects will issue drawings by way of architect's instruction and most will highlight any revisions or amendments on the new drawing issue. Nonetheless, the drawing must be scrutinised and checked to ensure that no other amendments have been made and that no further architect's instructions are required.

The contractor's quantity surveyor should also carry out checks on specification and quantities included within the bills of quantities, taking off quantities from drawings or checking measurements on site, and examining the contract bills to ensure that all works have been measured in accordance with the measurement rules which, in the case of the JCT 1998, is the Standard Method of Measurement of Building Works, 7th Edition, published by the Royal Institution of Chartered Surveyors and the Building Employers Confederation.

Another method is to take the contract bills and drawings to site and check them against works carried out. This particular method is very effective with housing schemes, either new-build units or modernisation schemes. Most multi-unit contracts will be measured by the client's quantity surveyor on a single unit basis, each house type being measured separately and totals being transferred to separate summary pages. Each unit can be checked in a relatively short space of time, simply by scrolling through the bills of quantities, checking each item in turn, examining the content of each item to ensure that it fully describes the item of work to be carried out and that the item of work has been measured correctly in accordance with the measurement rules. Any items not adequately measured in either quantum or description should be recorded and notified to the architect with a request that an architect's instruction is issued to cover any shortfalls or omissions.

On refurbishment projects, with complex multiwork content descriptions, a simple test is to analyse the item and see whether it can be drawn from the description. If not, is the item short of certain elements and is an architect's instruction required?

The contractor's quantity surveyor should check through both the invoice files and material order files on a regular basis, as these can be an 'Aladdin's cave' in discovering additional works. Materials can be abstracted and analysed against the bills of quantities, checking specification and total quantities against measured quantities once an adjustment has been made for waste, etc. Similarly, a regular check through the operative's time sheets can highlight where additional works have been carried out: some items will undoubtedly be in respect of rectification of work incorrectly carried out by the contractor, but others will relate to additional works which will require an architect's instruction to be issued.

At the end of any project or any work section the contractor's quantity surveyor should also walk through the job with the general foreman and the contracts manager. This occasion should be used to discuss the job in detail: the contractor's quantity surveyor can explain where he considers there have been areas of additional work and similarly the site team can report any areas where they consider variations have occurred to ensure that all additional works have been accounted for.

Remeasurement of Provisional Items

Clause 2 of JCT 1998, Contractor's Obligations Private with Quantities, requires under condition 2.2.2.1 that the contract bills, unless otherwise specifically stated, are to have been prepared in accordance with the Standard Method of Measurement of Building Works, 7th Edition. Within this document condition 1.1 requires that the bills of quantities should fully describe and accurately represent the quantity and quality of the works to be carried out. That being the case, there should be no need to include any allowance for provisional items or quantities. In practice, however, there are very few projects where this scenario can be the rule.

Very often the sections of the bills of quantities relating to substructures, drainage and external works will be marked in their entirety as provisional and will require full remeasurement once the final design is established and ground conditions are known. In some cases the bills of quantities will include provisional quantities only against items such as excavating and filling in soft spots, or a provisional quantity may be inserted to cover the possibility that some excavation may occur in rock. The client's quantity surveyor should include the item not only to cover the possibility of coming across such a situation on site, but also to ensure that should the works be required then there is already a contract rate against which to value the work.

More often than not, however, it will be necessary to remeasure such works on site as they are constructed. A point worth remembering is that the contractor is only entitled to be paid for work carried out in accordance with the contract drawings and measured strictly in accordance with the rules set out in the Standard Method of Measurement of Building Works, 7th Edition. For example, if the contract drawings indicate a 900 mm wide by 450 mm deep concrete foundation within the substructures, but the contractor constructs a wider or deeper foundation, then he is only entitled to include within his remeasurement a foundation width and depth as indicated on the contract drawings. The fact that the works are marked as provisional and require remeasurement does not alter the contract measurement rules.

In some circumstances it may be that additional work is required over net quantities. For example, soft spots could be encountered during excavation works, in which case the contractor, subject to agreement of quantity with the client's representative, will be entitled to be paid in full for these works carried out. Areas such as these require adjustment within the provisional quantities remeasurement section of the interim valuation or final account.

It is the responsibility of the client's quantity surveyor, as with other variations, to record and value these provisional works. Clause 13.6 gives the provision for the contractor to be present when measurements are taken and for the contractor to take any notes and measurements that they consider appropriate, a requirement that the contractor should equally afford to his subcontractors. In general, the contractor's quantity surveyor will record any such measurements as the works proceeds, agreeing the records with the client's representative. If, however, the quantity surveyors cannot attend site on a regular enough basis then they should arrange for agreed measurements to be taken, perhaps between the site foreman and clerk of works before the works are covered. From the viewpoint of the contractor's quantity surveyor, measurement without verification with the client's representative is a fairly pointless exercise if the works are to be covered after measurement. Photographs will assist with these records, but agreed records are preferable.

Provisional work is not restricted to ground work; builders' work in connection with mechanical and electrical installations is very often measured as provisional quantities, as are works in connection with external services. On refurbishment contracts there will be many unknown or unquantifiable elements at the time of bill production. Some projects will require complete remeasurement throughout the currency of works on site. Such contracts may well be let using a JCT contract with approximate quantities.

Whatever the remeasurement requirements of the project, constant monitoring and agreement are required and it is essential that measurements are taken as works are completed. How, for example, can brickwork repairs be measured once plasterwork has been completed, and how can the plaster repairs be measured once decoration has taken place?

Provisional works are not just restricted to those items that cannot be accurately measured in terms of quantity. The exact specification or the precise nature of some items may not be apparent when the contract documentation is being prepared.

Valuation of Nominated Subcontractors/Supplies

The valuation of subcontractors nominated under clause 35 of JCT 1998 is very much outside the duties of the contractor's quantity surveyor, but knowledge of the payment procedures or, more importantly, knowledge of the implications of non-payment is essential to the contractor's quantity surveyor so that they can keep management advised of the implications of withholding or delaying payment.

An interim certificate issued by the architect requiring payments to be made to nominated subcontractors will detail out and direct the contractor as to the amounts included within the interim certificate in respect of works carried by the nominated subcontractors. At the same time as the contractor receives the notification, the architect must equally inform the nominated subcontractors of the amounts that have been included within the interim certificate on their behalf.

Once amounts are included within an architect's interim certificate the contractor's quantity surveyor must ensure that payment is made to the nominated subcontractors within 17 days of the date of issue of the architect's certificate. The contractor's quantity surveyor should remember that if amendments have been made to the main contract which extend the payment terms, for example, from 14 days to 21 days, then an appropriate amendment must also be agreed with the nominated subcontractors and amendments made within the conditions of the nominated subcontract.

Once initial payments have been made to a nominated subcontractor the contractor's quantity surveyor will be required to provide evidence to the architect or his quantity surveyor that payments have been made in accordance with the contract provisions. It is important, therefore, that receipt of payment is requested and formally recorded.

Failure to pay the nominated subcontractor on schedule or provide proper evidence to that effect can result in the architect invoking the direct payment procedure. This procedure must begin with the architect certifying to the contractor that they have failed to make payment. A copy of this certificate must also be sent to the nominated subcontractor. Once these notices have been issued the architect is empowered to deduct from the next payment to the contractor the amount that the contractor should have paid to the nominated subcontractor.

Materials On Site

The JCT contract also allows for the inclusion within the interim valuation of materials on site (clause 30.2.1.2). Provided they meet fully with the required criteria, this means in effect that they are not brought to site prematurely and that they are adequately stored and protected from the weather before their incorporation into the works.

Earlier in this chapter the implications of delivery of materials to site were discussed in relation to keeping the contractor's cash flow healthy, and the contractor's project team must consider the implications of premature delivery in their deliberations on cash flow. From the contractor's point of view, there are many mitigating circumstances when it may be necessary to draw or call off materials earlier than when required, on the face of it. Certain materials have long delivery dates and will require early orders, which may result in delivery earlier than anticipated. It would be a foolhardy contractor who delays or suspends delivery of such materials. Similarly, to obtain certain materials the contractor may have to book manufacturing time and again it may be necessary to place orders and accept early supply; late delivery is simply not an option. Materials may have to be purchased in full loads even if their incorporation into the works is only required on an intermittent basis; for example, some precast concrete materials such as kerbs or drainage materials may be required at the onset of the project, while the remainder of the order is not required until the closing stages. The materials will not deteriorate and provided they are stored correctly the client's quantity surveyor should not raise any objection to their inclusion in the interim valuation. The correct storage of materials by the project team should be second nature, and the aim of reducing waste on site should more than adequately deal with the restrictions of the contract provision for materials on site.

If the contractor's team suspects that the architect may reject any unfixed material to be included within an interim valuation on the basis of premature delivery, early discussion with the architect and client's quantity surveyor, where the contractor can explain his reasoning behind the deliveries, may avoid unnecessary discord during the valuation process.

To value this particular section requires therefore a physical check and scheduling of materials on site based on quantity and value. It may be possible to obtain a schedule completed by the site manager, perhaps completed in conjunction with the clerk of works. It remains the responsibility of the contractor's quantity surveyor to ensure that the schedule produced is a full and accurate assessment of the unfixed materials on site. The importance of this schedule should not be overlooked or underestimated; depending on the project size the schedule of unfixed materials can amount to a considerable percentage of any interim valuation figure. Again, as with the measured work section, given that in-depth cost analysis will be carried out on each interim valuation, detailed scheduling is required. It is similarly unacceptable to schedule only the main materials; the whole site should be searched for unfixed material, including all sundry elements.

Most contractors will produce, on a weekly basis for each project, an accurate schedule of materials and goods delivered to site during the previous week. These schedules are used by the contractor's accounts and costing departments in the course of their administrative duties.

The contractor's quantity surveyor will find it helpful to reconcile materials delivered in total from material received sheets, deducting what is fixed and comparing the result with the physical measure on site. If in doubt as to the material being counted, he should take what references are available from the packaging or make a sketch to confirm its exact identity later. Material on-site schedules should always be discussed with the site manager, as specification changes may render some materials unusable and therefore they should be excluded from material on-site schedules. If there has been a specification change, then the value of residual stock should be included in assessment of the variation covering the specification change. A material on-site schedule should also include all materials belonging to subcontractors, provided that they meet the required criteria, as above.

Using this method of analysing material from material received sheets to assess materials on site, the contractor's quantity surveyor can produce a running abstract of the materials delivered to site and then deduct from these figures the amount of material that has been incorporated into the works. This method has its drawbacks: it takes no account of excessive waste or theft which can occur on site and could result in incorrect figures being used within the interim valuation. The abstract can, however, provide a useful record which can be used as a checking mechanism against actual quantity measured on site.

In valuing the unfixed material the contract allows for the 'total value' to be included within the interim certificate. The client's quantity surveyor will take particular notice of this provision, particularly if an item is incorrectly priced within the bills of quantities. Many client's quantity surveyors will attempt to reduce the 'value' of such material on the basis that the contractor cannot expect to be paid more than the allowance within the bill of quantities item/rate. While this methodology is not in accord with the contract provisions, in practice, if full value were allowed within the interim valuation, the contractor's quantity surveyor would have to account for the overvaluation within his valuation reconciliation.

Materials Off Site

JCT 1998 alters previous provisions relating to the payment for materials off site. In previous editions payment for materials held off site was at the sole discretion of the architect, provided that the strict criteria of the contract conditions had been met.

JCT 1998 condition 30.3 refers to 'the listed items', materials or goods or items prefabricated for inclusion in the works. These materials are listed by the employer and annexed to the contract bills. The amounts that would therefore be included in an interim valuation would be those included on the aforementioned list. Again, as with the previous editions, strict criteria must be met. JCT 1998 emphasises title or ownership of the material, requiring that the contractor provides reasonable proof that the ownership of the uniquely identified listed items is vested with them and that upon payment of such sums included within an interim certificate the materials become the property of the employer.

In addition to the question of ownership other criteria have to be satisfied. The materials must be:

- in accordance with the contract
- set apart from other materials, or clearly and visibly marked with a predetermined code
- marked with the employer's name and with the name of whose order they are held under and their destination, i.e. the works.

The contractor must also provide reasonable proof that the listed items are insured against loss or damage for their full value.

Retention

Clause 30.4 describes the retention that the employer can deduct from the interim valuation, details of which are indicated in the appendix to the conditions of contract. The standard contract provisions include the provision of 5% retention as a general rule, with the proviso that this should apply unless a lower rate is agreed. The contract also includes a footnote that where tender stage estimates the contract sum to be over £500,000, the retention percentage should be not more than 3%.

Thorough inspection of the contract conditions before tender submission should highlight to the management team whether unfair burdens are being suggested within the tender documentation in respect of the retention fund. Should that be the situation, the contractor may wish to make representation to the client's consultants before tender submission.

Clause 30.4.1.3 describes how the retention is reduced upon the issue of the practical completion certificate.

The contractor's quantity surveyor must keep in mind the appropriate dates when retention percentages should be reduced. Practical completion or sectional completion dates and the issue of the certificate(s) of making good defects are the key dates that trigger reductions in the retention fund. Failure to be diligent in monitoring and collecting outstanding retention monies will soon result in a downturn in the contractor's cash-flow situation.

A simple method of doing this is for the contractor's quantity surveyor to schedule key dates, i.e. date of project commencement, interim valuation dates and completion date, which should be the date for practical completion, and anticipated date for making good any defects, into both the front of the valuation file and their diary as a memory aid for future action. The above dates will be more complex if the sectional completion supplement forms part of the conditions of contract, but the principle of recording the dates remains the same. This monitoring procedure can also be used to prompt the contracts management into the timely completion of projects and of making good defects.

Liquidated and Ascertained Damages

Liquidated and ascertained damages are dealt with in clause 24 of the JCT form of contract. They are triggered following certification by the architect that the contractor has failed to complete the works by the date for completion stated in the appendices or any revised date for completion calculated after the architect has issued an extension of time to the contract period. The 'certificate of non-completion' is a *condition precedent* to the employer's right to withhold monies from an amount due on an interim certificate. Should the employer deduct such sums without the architect certifying to that effect, then the contractor can take whatever recourse the condition of contract allows which, in the case of JCT 1998, is adjudication or arbitration. Furthermore, if a new date for completion is fixed by the architect after issuing a non-completion certificate, then the effects of that initial non-completion certificate in allowing the deduction of liquidated damages by the employer are invalidated. Without further certification by the architect, the employer has no contractual right to withhold monies from the contractor and should return any monies previously withheld.

The amount of damages withheld can therefore be very changeable, as extensions of time are agreed and issued by the architect and, where applicable, the architect issues further certificates of non-completion. This difference in the amount of damages may change many times during the currency of a project, but will be finally resolved once the architect has completed his final review of the date of completion. This must take place 12 weeks after the date of practical completion, provided that the contractor agrees with the architect's assessment.

While the architect or quantity surveyor will calculate the amount of the liquidated damages due as a result of any such default on the part of the contractor, they are not empowered to make any adjustment to the amounts shown as due on valuations or interim certificates. The deduction of monies in respect of liquidated damages is not mandatory, it is solely at the discretion of the employer.

The liquidated and ascertained damages, as the name suggests, are a predetermined fixed estimate of the losses that the employer would suffer if the date for completion was not achieved by the contractor. The damages should include such items as loss of rent, loss of use and the direct costs arising out of such loss, such as interest on capital involved.

Other Deductions

Two further deductions can be made from amounts due under an interim certificate. If the contractor has not made timely payments to nominated subcontractors of amounts included in previous interim certificates, then the conditions of contract allow for the deduction of such equivalent sums to be deducted from future amounts due to the contractor.

The second of these allowable deductions is that of materials or workmanship that are not in accordance with the conditions of contract. Such elements are not strictly deductions, but rather omissions from the assessments made by the client's quantity surveyor. Many quantity surveyors will issue a statement with their interim valuations prompting the architect to consider whether all works included within the quantity surveyor's valuation comply with the above requirements. If that is not the case, then the architect could make an adjustment to the amount included in the interim certificate until the errors are corrected.

Value Added Tax (VAT)

VAT is dealt with as a supplemental provision (the VAT agreement) within JCT 1998 and requires that the employer pays the contractor on the supply of goods and services under the contract any tax properly chargeable in accordance with HM Customs and Excise Notice 708, VAT Buildings and Construction.

Part A of the notice sets out which goods and services can be zero rated. This section is applicable to contractors who are working on:

- new dwellings
- new relevant residential buildings
- new relevant charitable buildings
- a conversion for a registered housing association
- 'protected' buildings.

Part B of the notice deals with those involved in speculative building and will not be detailed here, except to say that work to new dwellings is ordinarily zero rated for the purpose of VAT.

Zero rating of VAT may apply in the following situations.

- The construction of a new dwelling or dwellings, which can be detached, semi-detached, a terrace of houses, a bungalow or a block of flats. This will include any necessary civil engineering work required within the curtilage of the site or development, such as roads and sewers, and the provision of services, i.e. mains water and sewerage, gas and electricity supplies to the nearest connection point. To qualify as zero rated the dwelling must be self-contained, with no direct access to any other dwelling or part of a dwelling, it must have been granted statutory planning consent and be built in accordance with that consent, and the separate use of the dwelling must not be restricted by the terms of any covenant, statutory requirement or similar provision.

- The construction of an independent annexe for charitable purposes built onto an existing structure. This criterion does not include small additional living accommodation built onto or in the grounds of an existing house. These structures, commonly known as 'granny annexes', will be standard rated even if the new structure is entirely separate from the original structure.
- The conversion of a building from non-residential into residential use by a relevant housing association, but only where some of the facilities are shared by residents, e.g. cooking and dining. To qualify within this section the residential building must be:
  - a children's home
  - an old people's home
  - a home for the disabled (where care is provided)
  - a home providing care for persons who suffer or have suffered from drug or alcohol dependency or mental disorder
  - a hospice
  - residential accommodation for students or school pupils (including dining rooms and kitchens if used predominantly by living-in students)
  - residential accommodation for the armed forces, such as unaccompanied officer's mess quarters and most barrack blocks
  - living accommodation for members of religious communities
  - any other communal living accommodation which is the main or sole residence for at least 90% of the residents.
- Specifically excluded from this zero rating are hospitals (private or NHS), prisons and other penal institutions, and hotels, inns or similar establishments. An additional requirement of this sector is that before the supply is connected the end user must give a relevant certificate to the contractor confirming that the building will be used for qualifying purposes.
- Enlarging or extending a building where the enlargement or extension creates an additional new dwelling and where the separate disposal is not prevented by any planning or similar consent.
- Civil engineering work for development of a permanent residential caravan park.
- Where approved alterations are being made to a listed building which is a dwelling, a relevant residential or relevant charitable building, or will become so as a result of them.

With the exception of those matters scheduled above, the basic position is that all goods and services are standard rated. This includes the construction of new buildings not in any of the above categories, all repairs and maintenance work, and any work done to an existing building, including any alteration, extension, reconstruction, enlargement or annexation of an existing building, except as specifically excluded in the above section.

The above represents a summary of the matters referred to within VAT Notice 708 and is intended as a guide only. For specific queries the contractor's quantity surveyor should examine in detail Notice 708 and if any doubt arises from these investigations then discussions should take place with the local office of HM Customs and Excise, where a ruling can be established.

To comply with the provisions of the VAT agreement within the contract conditions the contractor should, not later than 7 days before the date of the first interim valuation, give notice to the employer, with a copy of the notice going to the architect, of the rate of tax chargeable on the supply of goods and services for which interim certificates and the final certificate are to be issued. A further notice should also be sent to both employer and architect should the rate of VAT change under statute.

At the time of making each interim valuation the contractor's quantity surveyor must ensure that the appropriate application for VAT is made. In general, this will take the format of a provisional assessment of VAT for each interim valuation with the final assessment being made once the final account is agreed. The assessment should identify separately the net amounts of the interim certificate to be zero rated and those amounts that attract VAT at the standard rate, and also the rate of VAT to be charged, and must also state the grounds on which the contractor considers such supplies to be chargeable.

Provided the employer has no objection to the contractor's assessment, then the amounts so calculated should be added to the amount of the interim certificate and paid to the contractor within the period stated in the contract provisions. Once this payment is made then the contractor must issue to the employer a receipt of payment as a tax invoice to cover the payment made.

If during the currency of the contract an overpayment is made in respect of VAT paid to the contractor then this can simply be redressed once the final assessment is made, with any overpayment being refunded to the employer within 28 days of the employer's notice of such an overpayment.

Condition 2.2 within the VAT agreement deals specifically with the provision of VAT on deductions made by the employer in respect of liquidated and ascertained damages. The condition states that payments of VAT to the contractor should be made in full as though no deduction had been made.

The schedule in Figure 3.6 represents a standard format which the contractor's quantity surveyor can use when making a provisional assessment of VAT.

Site Visits

The number of visits necessary to any site will depend on many factors, including the size of the project, the number of variations and the type of project. Many modernisation or revitalisation schemes where most of the trades require remeasurement will require regular, if not constant, on-site work, whereas on small, less complicated projects the monthly valuation visit may be sufficient. Equally, on large projects a team of quantity surveyors may be required to cover effectively the quantity surveying duties on the contractor's behalf.

However, the more often site visits are made by the contractor's quantity surveyor, the greater the job knowledge and the less will be missed, thus maximising the return from the project.

Photographs, particularly dated photographs, can be of immense value not just for valuations but also as a permanent record for possible later use in settling the final account or in claim situations. It is also worthwhile recording photographically any variation work that takes place and requires remeasurement, particularly those areas of work that will subsequently be covered up or buried.

Talking to the operatives, not just passing the time of day, although this can be valuable, will be as rewarding as photographs. Since it is the operatives who

Commercial management in construction

PROVISIONAL ASSESSMENT	By the Contractor by the VAT agreement
IMPORTANT	This is not a tax invoice and cannot be used to validate a claim for input credit.
CONTRACTOR	MURRIBELL BUILDERS LTD
CONTRACTORS ADDRESS	HIGH STREET SUNNISIDE ENGLAND E1 2BB
EMPLOYER	_____
CONTRACT WORKS	_____

NET AMOUNT of VALUATION £_____ (VAT EXCLUSIVE) Valuation No.____

Provisional assessment of respective values of supplies of goods and Services for which the above Valuation is to be issued to the Architect which are chargable on the Contractor at the relevant time of supply under Regulation 1-15 of the Value Added Tax (General) Regulations 1995.

Category (I)	Zero Rated	@0%	£
Category (ii)	Positively Rated	@17.5%	£ _____
		Total £	_____

Ground on which Contractor considers the suppliers included in Category (ii) are chargeable at a positive rate of tax.

Work in repairs and maintenance liable to VAT at the standard rate.

Signed _____   Date _____
Murribell Builders Ltd (Contractor)

Figure 3.6 Provisional assessment form.

actually carry out the work, they will know the complexity of the work, which may equally turn out to be a variation. The need for liaison between site personnel and the contractor's quantity surveyor cannot be overemphasised: it is vital that a dialogue is established and maintained to enable a free flow of information between disciplines. An 'us and them' attitude will not work.

Checking the time sheets, particularly with the project manager, can also prove worthwhile. Much information can be extracted from these records and when variations are being calculated, particularly those to be valued at a fair rate, the evidence will be at hand to be used as necessary in the compilation of the rate. These records must always be stored in a safe place and not left on site, as they may be required at a later date for proving variation costings.

On larger projects site meetings will be held on a regular basis, in general monthly, but on fast-track, high-value projects they may be held fortnightly or even weekly. Again, the records that ensue from these meetings can provide valuable information for the contractor's quantity surveyor to analyse and use.

These meetings are an excellent forum for extending project knowledge. Many matters will be discussed and it is a good opportunity to put forward any requests for variations that have not been picked up in the normal process. All parties are generally represented, including the architect, quantity surveyor and clerk of works, as well as the contractor, so that any discussions regarding the validity of the variation requests can take place among all concerned. The added advantage is that the site meetings are formally minuted, thus providing a record of variation requests for future monitoring.

When measuring on site for variation work it is imperative that records are fully headed with the architect's instruction number if known, or the clerk of works' direction number, a drawing reference, the description and nature of the variation, the date and those present during the measuring process. Each variation should be separately recorded and measured in full, i.e. full addition and omissions; measuring the net effect of variations is not sufficient. The end result of this site work should be recorded as detailed earlier in this section.

One final word on site visits: **safety**. It is *absolutely imperative* that every care is taken when on site. Construction sites are extremely dangerous places, and hard hats and correct safety footwear must be worn at all times. One should never walk on a board lying loose on site, as it may cover a hole or a void, never climb on loose material, and exercise extreme caution when walking along scaffolding, watching for loose boards, missing hand rails, poorly restrained ladders, etc. If any faults are evident, they must be reported immediately to the site management and the scaffolding must not be used.

Final Accounting

Completion of the final account on any project requires a joint and sustained effort from all parties involved. This is not just a job for the architect or the client's quantity surveyor, but requires considerable input from the contractor's quantity surveyor, input that needs to be proactive and not reactive.

JCT 1998 clause 30.6 sets out the requirements relating to the preparation of the final account:

> Not later than 6 months after Practical Completion of the Works the Contractor shall provide the Architect, or, if so instructed by the Architect, the Quantity Surveyor, with all documents necessary for the purposes of the adjustment of the Contract Sum including all documents relating to the accounts of Nominated Sub-Contractors and Nominated Suppliers.

This contractual requirement for the contractor to provide 'all documents necessary' will in some instances create tension between parties involved in the settlement of the account. Undoubtedly, requests will be made for information that the contractor may regard as privy to them, but provided a reasonable approach is taken by all parties, common sense should prevail. It is important for all parties to recognise the needs of others; for example, the contractor must understand that the final account will generally be audited by a third party who may have no particular knowledge of the construction process or of the particular project being assessed. The auditor will look for documentary evidence that the contract provisions have been strictly adhered to and that demonstrable evidence of costings is available for verification. An example could be a claim made by the contractor for payment for additional works that have been carried out, but without an official architect's instruction being issued. The contractor could provide all necessary supporting evidence of costs, etc., but without the architect's input in bringing the architect's instructions up to date during the completion of the final accounting process an auditor would disallow the contractor's claim.

Clause 30.6 sets a further time scale for completion of the final account: the architect or the quantity surveyor should issue to the contractor the final statement of adjustments to the contract sum, and also any ascertainment of loss and expense, not later than 3 months after receipt of the information from the contractor.

Effectively, clause 30.6 allows a maximum period of nine months for completion of the final adjustment of the contract sum.

In general, however, it is of benefit to all parties to agree accounts as expeditiously as possible, and provided the variation account is kept up to date and areas of change are agreed as the project proceeds it is possible to reduce substantially this period for agreement.

By being proactive in completion of the final account the contractor's quantity surveyor should avoid shocks to the other team members. Early warning of the costing and time implications of additional items of work will allow the team to better manage the overall effects arising out of any variations.

It is important to the contractor that all works carried out are properly measured and valued, and all costs incurred by the contractor need to be assessed by their quantity surveyor before agreeing the account with the client's representative. However, all parties will benefit from early settlement. The employers will know their final level of commitment to the project, and the contractors, architects and quantity surveyors will be free to move on to new projects having to revisit the current scheme, save for the issue of the final certificate after completion of the making good of defects.

During this period both quantity surveyors will have subcontract accounts to settle. The client's quantity surveyor will need to agree any nominated subcontractors or suppliers and the contractor's quantity surveyor will have a myriad of domestic subcontractors to settle. As with the variation account, as subcontract works are fully completed during the contract period, so their accounts can be agreed at that time and incorporated into a draft final account document. The final account should be seen not as a document wholly produced at the end of a project, but as a continuous reassessment of the contract sum as events dictate. As with the main account, agreeing as many areas as possible during the contract period will pay dividends at the end.

The finalisation of the subcontract accounts needs to be managed in tandem with the agreement of the main project account with the client's quantity surveyor. All additional works submitted by the subcontractors must be assessed and included within the main account, or they must accept and agree why certain areas of their submissions are to be excluded from the account. As with the main account, as areas are agreed with subcontractors they should be signed off to avoid

Interim valuations 53

MURRIBELL BUILDERS LTD
High Street
Sunniside

SUBCONTRACTOR
(Name & Address)

FINAL ACCOUNT STATEMENT

CONTRACT _____

CONTRACT No: _____

Amount of completed works executed .... £ _____
Less ____ % Main Contractors Discount

We the undersigned, agree that the above amount represents a full and final total value for works executed by ourselves at the above Contract and that on receipt of the outstanding balance will have no further claim or claims against Murribell Builders Ltd.

SIGNED _____
FOR   _____

_____
_____   DATED _____

Figure 3.7 Final account statement.

the need to revisit. Finally, once the whole account has been settled a statement to that effect should be signed by the subcontractor. The subcontract final account statement sheet (Figure 3.7) is suitable for this purpose.

The contractor's quantity surveyor should also keep a running schedule of the status of subcontract accounts as they are agreed. The contract analysis schedule (Figure 3.8) can be used to achieve this. As the schedule indicates, all subcontract types can be summarised using this document. In addition, sundry costs that are not allocated to the normal labour, plant and material sections, such as statutory authority fees or consultants' fees, can be identified. As subcontractors are agreed, they can be colour coded to indicate that their accounts have been finalised.

With regard to the main project account and its agreement with the client's quantity surveyor, clause 30.6.2 sets out how the contract sum should be adjusted. The original contract sum is adjusted in line with the contract provisions, starting with the omissions for:

- all prime cost sums and provisional sums, including the value of work relating to items included as approximate quantities

54 Commercial management in construction

Murribell Builders Ltd

CONTRACT ANALYSIS

Project		Project Nr.		Date	
Anticipated/Agreed Final Account Amount			£0.00		
Sundry Invoices			£0.00		

Nominated Subcontractors/Suppliers - Statutory Authorities

Name	Amount of Subcontract Liability	Subcontract Net Liability	Discount %	Amount

Domestic Subcontractors

Name	Amount of Subcontract Liability	Subcontract Net Liability	Discount %	Amount

Labour Only Subcontractors

Summary
Nominated
Labour Only
Domestic   0.00   0.00   0.00   0.00   0.00   0.00

Making Good Defects Allowance   £0.00

Figure 3.8

- the valuation of any work omitted resulting from the issue of an architect's instruction
- deductions that may be necessary following the default of the contractor with regard to accepted errors in setting out, materials or goods that are not in

accordance with the contract and the potential costs of making good defective works properly scheduled at the end of the making good of defects period, but by acceptance of the employer not carried out provided a monetary consideration is made
- the amount of fluctuation allowable to the employer as allowed in clause 38, 39 or 40
- any adjustment necessary resulting from the determination of a nominated subcontractor where renomination is necessary under the provisions of clause 35.24.

Once the omissions have been accounted for, the final account should be adjusted to include for areas of addition:

- the finally agreed amounts in respect of nominated subcontractors and suppliers, including any work carried out by the contractor in accordance with clause 35.2, incorporating any appropriate discounts and profit allowable by the contract
- any fees or charges, the costs associated with opening up for inspection, patent right and the cost of 21.2.1 insurances as are allowable within clauses 6.2, 8.3, 9.2 and 21.2.3, respectively
- the valuation of variations issued by the architect
- the valuation of works carried out in respect of the expenditure of provisional sums and the valuation of work covered by approximate quantities
- any amounts ascertained in respect of loss and expense under clause 26.1
- any amounts due to the contractor in respect of insurance matters under clause 22B or C
- the amount of fluctuation allowable to the contractor as permitted by clause 38, 39 or 40.

While the above is not a definitive schedule of those matters that require adjustment in the final accounting process, the schedule indicates areas that may require adjustment during the course of normal construction projects. Equally, as in many areas of the construction process, each project must be examined on its own merits and an appropriate format established and agreed between parties.

Once the final account has been fully agreed it is important to obtain formal agreement to that effect. As with subcontractors, a statement of account should be issued which sets out fully the amount of the agreed final account. If the contract stipulations regarding the issuing of the final certificate preclude its issue at that point, but it is important that the contractor's quantity surveyor ensures that a further interim certificate is issued in good time to facilitate payment of any outstanding balances. In requesting such further payment the contractor's quantity surveyor must have regard to the restrictions on the issuing of further interim certification, in particular the final certificate, as set out within clause 30.7. This condition stipulates that any further certification cannot be made within 28 days of the proposed issuing of the final certificate. The contractor's quantity surveyor must therefore make a judgement as to whether it is better to wait for the final certificate to be issued, when all outstanding monies will be released including any necessary retention releases, or to have outstanding final balances certified and wait a further 28 days for the outstanding retention monies.

The contract requirements regarding the issuing of the final certificate referred to above are covered in clause 30.8 of JCT 1998. The final certificate must be issued not later than 2 months after whichever of the following requirements occurs last:

- the end of the defects liability period
- the date of the issue of the certificate of making good defects
- the date when the architect sends the final contract adjustments, i.e. the final account, to the contractor.

The effect of issuing the final certificate is discussed in Chapter 6.

Recommended Further Reading

Seeley, I.H. (1997) *Quantity Surveying Practice* (Second Edition). Macmillan.
Wainwright, W.H. (1967) *Variation and Final Account Procedure*. Hutchinson Technical Education.
Willis, C.J., Ashworth, A. and Willis, A. (1994) *Practice and Procedure for the Quantity Surveyor* (Tenth Edition). Blackwell.

4 Subcontractors

The procurement and financial management of subcontract works will involve a considerable amount of the contractor's quantity surveyor's normal timetable. Subcontract services in general can form the greater part of any construction project, with many contractors opting to subcontract the whole of the works apart from the general or project management services. This almost unilateral use of subcontract services is not just the result of the specialists' works required on many projects, but can also be used by contractors moving into new areas of operation, when the prudent option may be to pass the potential risk to others. In addition, many contractors enlist the services of labour-only or labour and plant-only subcontractors. Again, this procurement method can cut down the risk to the contractor with the build operation. This methodology means that the contractor has no obligations in respect of long-term employment of operatives and it also removes the abortive costs associated with such matters as non-productive overtime and those resulting from reduced production through inclement weather.

While the above practice, particularly in relation to labour-only subcontractors, can prove cost-effective in the short-term, contractors must be aware that many labour-only subcontractors do not operate apprentice schemes, a situation which will eventually lead to skill shortages within the industry. In addition, the contractor must pay particular attention to the quality of work produced by the subcontractors, as ultimately the contractor is fully responsible to the employer for all matters of quality, timing and cost on any project.

Essentially, there are only two types of subcontractor, client chosen and contractor chosen; however, subcontract packages, in general, fall into four categories:

- nominated subcontractors
- named subcontractors
- domestic subcontractors
- labour-only and labour and plant-only subcontractors.

Nominated Subcontractors

Nominated subcontractors are subcontractors where an architect has exercised his right under clause 35 of JCT 1998 to select specific subcontract services. This selection can be facilitated either by the architect singularly naming subcontractors within the original tender package to the contractor, or by including the provision of prime cost (PC) sums or provisional sums within the tender documents for later conversion into nominated subcontract services.

To nominate a subcontractor a strict methodology must be followed by all parties involved.

JCT 1998, clause 35.4, sets out the documents used within the process:

Name of document	Identification term
The Standard Form of Nominated Sub-Contract Tender	NSC/T
1998 Edition which comprises:	
Part 1: The Employer's Invitation to Tender to a Sub-Contractor	Part 1
Part 2: Tender by a Sub-Contract	Part 2
Part 3: Particular Conditions (to be agreed by a Contractor and a Sub-Contractor nominated under clause 35.6)	Part 3
The Standard Form of Articles of Nominated Sub-Contract Agreement between a Contractor and a Nominated Sub-Contractor, 1998 Edition	Agreement NSC/A
The Standard Conditions of Nominated Sub-Contract, 1998 Edition, incorporated by reference into Agreement NSC/A	Conditions NSC/C
The Standard Form of Employer/Nominated Sub-Contractor Agreement, 1998 Edition	Agreement NSC/W
The Standard Form of Nomination Instruction for a Sub-Contractor	Nomination NSC/N

The procedure for nomination using the above documentation is set out within the remainder of clause 35.

From the contractor's point of view, notification will start upon receipt of an instruction issued on Nomination NSC/N by the architect nominating the subcontractor. The contractor should at this juncture consider the timing of this instruction; if the nomination is given late by the architect, then there is an entitlement of extension of time within condition 25.4.6 together with any financial effect being catered for as loss and expense within condition 26.2.1.

The contractor can, using the provisions of clause 35.5, make *reasonable objection* to the architect's nomination. Any such objections must be in writing and should be made at the earliest practicable moment, but in any event not later than seven working days from receipt of the architect's nomination. These objections can vary considerably; for example, the contractor may have worked in the past with the proposed subcontractor and problems of quality or programming may have occurred, the subcontractor's valuation procedures may have proved ineffective, or the subcontractor may have resisted the teamwork approach. Provided the contractor's objections can be demonstrated to be reasonable, the architect would be well advised to consider the caution.

Before the architect issues the nomination, which can be accomplished either by negotiation or by competitive tender, NSC/T Part 1 will have been completed by the architect, and the subcontractor and employer will have completed and signed NSC/T Part 2. Both of these documents will form part of the documentation issued to the contractor with the preliminary notice of nomination. In addition to the above, the original tender documentation included with NSC/T Part 1 will be included with the nomination to the contractor, as will agreement NSC/W, also completed by the employer and the nominated subcontractor.

If the contractor so requests, the architect should give the contractor the opportunity to price works covered by provisional or prime cost sums. To facilitate this, the contractor should give early notice of their desire to price such elements. The contractor's request may, however, have been overtaken by events, in that the architect may have already sought quotations from other subcontractors, before the main tender documents were issued. In allowing the contractor to offer a quotation, the architect must, in fairness to other subcontractors, ensure that all parties are pricing on a level playing field. The architect would need to consider whether the contractor's price includes the appropriate discounts and also includes for profit and attendances.

Once the contractor has received the architect's instruction nominating the subcontractor they must complete NSC/T Part 3 in agreement with the nominated subcontractor, provided they do not object to the nomination. Both parties must sign this document as being agreed and also execute the agreement NSC/A with the nominated subcontractor.

Once the above documentation is completed the contractor should send a copy of the NSC/A to the architect.

The use of nominated subcontractors within building contracts is becoming less widespread. While the nomination procedure affords both contractor and subcontractor certain benefits – for example, both have considerable protection in the event of the default of the other, particularly in the event of the insolvency of the other – the benefits to the employer are limited. The employer will have the benefit of choosing who will carry out the work, and where appropriate they can draw on the subcontractor's design expertise. In general, specialist subcontracts such as mechanical and electrical engineering and lift installations have traditionally been engaged as nominated subcontractors. All three require considerable preinstallation input in terms of design or lead-in time before delivery, and early discussion or negotiations with the employer prior to nomination can avoid delays once the contractor has been appointed. From the nominated subcontractor's point of view, the major benefit is that of payment protection. If the contractor fails to discharge payments to the nominated subcontractor in accordance with condition 35.13.2, the nominated subcontractor is entitled to receive payment direct from the employer, who in turn will make a corresponding reduction in payments to the contractor (under condition 35.13.5 and NSC/W 7.1 – see Chapter 3).

The appointment of a nominated subcontractor brings with it certain risks to the employer. If the nominated subcontractor delays the works then the contractor will be entitled to an extension of time, denying the employer the right to recover liquidated and ascertained damages from the contractor. The employer can make a claim for damages for delays caused by the nominated subcontractor under the terms of NSC/W, but in the event of the insolvency of the subcontractor the employer will almost certainly lose out. In addition, given that the contractor has no vested interest in pushing the nominated subcontractor to complete, the employer may lose any impetus from the contractor to complete the works promptly, with the contractor relying on the default of the nominated subcontractor to cover their own inefficiencies.

From a contractor's point of view, the main disadvantages of nomination are that the contractor does not choose the subcontractor. The choice is almost entirely beyond the contractor's control, except that the contractor can make reasonable objections to the nomination. That aside, the contractor is required to use whichever subcontractor the architect nominates for the works. In addition, it is mandatory for the contractor to use the standard formal nomination process and documentation. This takes considerable time and effort to complete and the contractor cannot use their own standard terms and conditions when setting up the subcontract order.

While the contractor is responsible for the default of domestic subcontractors, clause 35 substantially diminishes the contractor's liability in regard to subcontract works. The employer's remedy for default of the nominated subcontractor lies against that nominated subcontractor under NSC/W, and not the contractor. The contractor also has the benefit that they are not responsible for the design of works included in the subcontract package (condition 35.21). In the event of default by the nominated subcontractor, the contractor will be entitled to an extension of time to cover any consequent delays. In addition, provided the appropriate notices are issued by the architect (condition 35.15), the contractor can make a claim against the nominated subcontractor for the resultant damages brought about by the nominated subcontractor's default.

In valuation terms, the contractor also has the advantage that in general the nominated subcontractor's accounts will be dealt with independently without the contractor's imput being required. The contractor is advised of the amount included in respect of works carried out by the nominated subcontractors during the valuation process.

To obtain payment the nominated subcontractor should first submit to the contractor his account or interim valuation for the works carried out to date. This submission should be made in sufficient time to allow the contractor's quantity surveyor time to forward the nominated subcontractor's application to the client's quantity surveyor for verification and authorisation during the normal valuation process. The contractor may consider allowing the nominated subcontractor to forward copies of their interim applications directly to the client's representative, to speed up the process of verification for valuation purposes. Once authorisation has been made by the client's quantity surveyor within the interim valuation, the architect will issue the interim certificate inclusive of such amounts due to the nominated subcontractors, and within the same process issue directions to the contractor as to the specific amounts included within the interim certificate in respect of works carried out by the nominated subcontractors. At the same time, and in addition to the directions issued to the contractor, the architect must notify the nominated subcontractors of the amounts included within the interim certificate in respect of their works.

Payment to nominated subcontractors is covered in Chapter 3, but to summarise, the contractor's quantity surveyor must keep in mind the specific requirements of the payment provisions in the contract documentation. If the contractor is in default on payment, they will run the risk of having future payments to the nominated subcontractor paid directly unless they can provide evidence that timely payments are being made. This situation would lead to the contractor losing the entitlement of discount. The contractor would be liable to pay interest at a rate of 5% over current base rate and would also be liable for the resultant loss and expense, if the nominated subcontractor, upon notice, suspended further performance of the works as a result of the non-payment.

The conditions of contract contain no provisions for the contractor to issue extensions of time to the nominated subcontractors without the written consent of the architect. If the nominated subcontractor fails to complete the subcontract works in accordance with the agreed periods allocated for their works, the contractor must notify the architect of these events. It is then the responsibility of the architect to consider the delay and apportion extensions of time appropriately. Once the architect has issued the appropriate extension of time notices to the contractor, the contractor can correspondingly issue such extensions to the nominated subcontractor.

If the works are delayed because of default on the part of the nominated subcontractor, as already indicated, the contractor is entitled to seek recovery of damages from the nominated subcontractor. As a condition precedent of making such recovery, the contractor must first give notice in writing to the architect that the nominated subcontractor is delaying the progress of the works and, secondly, must obtain a specific notice from the architect specifying the default on the part of the nominated subcontractor. If there is no certificate issued by the architect, then even if the nominated subcontractor is in delay the contractor cannot claim or offset loss or damage for the delay by the nominated subcontractor. This certificate is a *condition precedent* to the contractor making the claim and must have been issued before the date on which payment became due, i.e. at least 5 days before the final date for payment. In addition, the contractor must follow the strict criteria set out within the subcontract conditions (NSC/C) before making any deduction or offsetting from amounts due to the nominated subcontractor. The contractor cannot simply withhold monies from the nominated subcontractor without complying with these contractual requirements. These provisions require that the contractor must give notice in writing to the nominated subcontractor specifying his intention to offset monies from payments due. This notice must quantify in reasonable detail the amount of the proposed offsetting and the reason why the deduction is being made. The notice must be given not less than 3 days before the date of issue of the interim certificate and not later than 5 days before the final date for payment.

One final point on nominated subcontractors is discount. Under the provisions of the nomination the contractor is entitled to deduct from the certified amounts a discount of 2.5%. The contractor would lose this amount if payments were made directly by the architect as a result of the contractor's default in making previous payments promptly. Some contractors will, however, assume the benefits of this discount provision when assessing their original tender submission. If a tender includes the provision of PC sums, many contractors will assume that they will be expended in total and net their tender down by an amount equivalent to what the discount would have been. The contractor will then assess the tender in terms of what level of overheads and profit should be included, but based on figures that represent the net amount of nominated subcontract works. The basis of the calculation is that the contractor will be satisfied with the assessment for overheads and profit added to the project without the need for further enhancement from allowable discounts. This has the effect that the contractor's tender may be more competitive than that of their rivals in the tendering process. Using this methodology, however, problems will arise if the PC sums are not expended in their entirety, i.e. not just in terms of sums not expended at all, but also in respect of sums where individual amounts per subcontractor fall short of the original sums. In effect, this means that the contractor has made an adjustment within the tender for discount that does not exist, leaving a shortfall which will have to be adjusted against the original overhead and profit assessment.

Named Subcontractors

From a contractor's point of view, the singular naming of subcontractors is becoming more and more prevalent on many projects. This methodology is adopted by architects, who can obtain some of the benefits of nomination without all of the associated burdens identified above, but more importantly the contractor is required to take responsibility for the subcontractor's performance when the subcontractors are chosen by others.

In general, the use of named subcontractors is a provision found within the Intermediate Form of Contract 1998 (IFC 98), although there are provisions within other contracts which provide for a similar approach. Unlike JCT 1998, the IFC contains no provision for nomination. However, the provisions for naming a subcontractor under IFC 98 represent very similar conditions, albeit in a considerably diluted manner.

The key features of naming a subcontractor within the provisions of IFC 98 are as follows.

- The subcontractor is selected by the architect and named using one of two methods, being named either in the original main contract documents, or in an instruction to expend a provisional sum.
- The standard tendering procedure set out in clause 3.3.1 of IFC 98 provides for the use of standard form of tender NAM/T, and also requires that such procedures are dealt with within 21 days of the contractor entering into the main contract conditions.
- The use of standard named subcontract conditions, NAM/SC, is also mandatory.
- By an express provision in the main contract, the main contractor will not be liable for any design provided by the named subcontractor.
- The named subcontract work is priced by the main contractor at his own rates and prices and is not the subject of a PC sum.
- Once appointed, the named subcontractor is treated in a very similar manner to an ordinary domestic subcontractor, so the architect has no involvement in matters such as certifying payments to him or deciding upon his entitlement to extensions of time.
- The main contractor is not entitled to an extension of time resulting from delays caused by named subcontractors.

- In the event of the determination of the named subcontractor's employment, an extraordinary complex set of provisions will apply which result in the sharing of risk between the main contractor and the employer.

A further point worth noting is that unlike JCT 1998, IFC 98 contains no corresponding provision for the naming of suppliers.

As indicated above, the procedure for naming a subcontractor under the provisions of IFC 98 varies according to whether the subcontractor is being named in the main contract documents, which effectively means that the subcontractor is chosen before the contractor, or in an instruction to expend a provisional sum once the contractor has been appointed.

Precontract Procedure

- Initially, the architect completes the invitation to tender in Section 1 of NAM/T and sends it to those being invited to tender for the subcontract works, together with specifications, schedules of work or bills of quantities describing the works to be carried out, and associated drawings as appropriate.
- The tendering subcontractors then complete Section 2 of NAM/T, giving details of their tender, and return it to the architect.
- The architect will then, generally in consultation with the employer and their quantity surveyor, make the selection of the subcontractor to be named and countersign Section 2 of the NAM/T form submitted by the selected tenderer.
- When tenders are invited for the main contract work, tenderers should be provided with a description of the works to be executed by the named subcontractor, together with the completed Sections 1 and 2 of NAM/T.
- Within 21 days of entering into the main contract, the contractor is required by clause 3.3.1 of IFC 98 to execute the standard articles of agreement in Section 3 of NAM/T, which incorporates, by reference, the standard subcontract conditions of NAM/SC
- IFC 98 then requires that the contractor confirms to the architect the dates on which individual subcontracts were entered into.

A notable omission from this selection process is that the contractor has no express provision of reasonable objection to the chosen named subcontractor. The theory is that the contractor is aware at tender time of the particular named subcontractors to be used, and has the option at that juncture of either making any allowances within the tender for same or not tendering for the works at all.

Section 2 of NAM/T sets out not only the subcontractor's tender price but also the periods of time required within the contract programme for the subcontractor to carry out the subcontract works and the special attendances required from the contractor. Ordinarily these matters will be dealt with by negotiation between the contractor and the named subcontractor and generally do not present a problem in completing the subcontract paperwork.

If, however, the contractor is unable to agree terms with the named subcontractor, IFC 98 sets out the rules that must be followed in relation to such disagreements. In such an event, the contractor must immediately inform the architect of the problem and set out in detail why agreement cannot be reached.

Provided the architect is then satisfied as to the particulars of the discord, instructions should be issued accordingly to resolve the impasse. The architect may:

- change the particulars so as to remove the problem; for example, reconsider the procurement of an item of special attendance, perhaps making the subcontractor carry out the works within their subcontract, or where there is more generally a problem, amend the contract completion date so the named subcontractor's programme can be accommodated within the contractor's programme;
- omit the work in its entirety from the contractor's commitment and perhaps, if practicable, carry out the works as a separate contract with the named subcontractor after the contractor has completed the main works;
- omit the works from the contract and substitute a provisional sum with the option of naming another subcontractor to carry out the works.

Whichever option is taken by the architect the variation must take the form of an architect's instruction so that the contractor can recover any costs so incurred as a result of the amendments and so that appropriate extensions of time and associated loss and/or expense can be considered should they be required.

Postcontract Procedure

This section deals with the procurement of named subcontractors during the currency of the contract, i.e. where the architect issues instructions naming a subcontractor within the expenditure of a provisional sum. While the initial procedure is similar to that of precontract procurement in that Sections 1, 2 and 3 of NAM/T still require completion, the main difference is that the contractor now has the right of reasonable objection. Any such objection needs to be made within 14 days of the issuing of the architect's instruction; if, however, no objection is made the contractor must then complete Section 3 with the named subcontractor.

This particular version of the naming procedure does not provide a mechanism for resolution if the contracting parties cannot agree terms, presumably relying on the contractor's right to reasonable objection to resolve any problems, but as with the previous procurement route, the architect can deal with any amendments required by issuing the appropriate instructions.

The use of named subcontractors means that the employer can still control who carries out the works. It also allows, in a similar fashion to nomination, for early appointment of subcontractors. Again, as with nomination, through the use of ESA/1 the employer has a direct link with the subcontractor in respect of design. The greatest advantage to the employer is, as pointed out earlier in this chapter, that the employer has no responsibility for the performance of the chosen subcontractor, unless the default or insolvency causes the determination of the named subcontractor's employment; in such situations the major burden of costs associated with such determination rests with the employer. Contractors need to pay particular attention to this provision. The requirement for renaming subcontractors must be strictly adhered to and affords the contractor considerable protection in the event of such default by the named subcontractor.

From the contractor's point of view, the downside, as with nomination, is that the choice of subcontractor rests elsewhere, as does the choice of subcontract conditions. Most contractors will have their own conditions of subcontract, but these cannot be used in situations where the architect names a subcontractor, when it is mandatory to use NAM/SC.

The procurement of subcontractors, whether by nomination or by the singular naming of subcontractors, has the added advantage to the subcontractors that the contractors cannot embark on the 'dutch action' approach to selection, with tenders being obtained very much as they would be for the main contract works.

Domestic Subcontractors

Domestic subcontractors are those subcontractors whose selection rests generally with the contractor. Both JCT 1998 (clause 19) and IFC 98 (clause 3.2) recognise that the contractor may wish to sublet part of the works. Clause 19 describes a 'domestic subcontractor' as a person to whom the contractor sublets any portion of the works, other than a nominated subcontractor. Both contracts require that the contractor should not sublet any part of the works, other than by nomination or naming procedures, without the written consent of the architect or contract administrator. Both contracts also include the provision that the architect or contract administrator shall not unreasonably delay or withhold such consent as to the contractor's choice of subcontractor or which areas of the project will be sublet.

With domestic subcontractors, the contractor remains wholly responsible for carrying out and completing the works. The contractor is fully responsible for the performance of the subcontractor selected and bears all of the risk as though they were carrying out the works themselves.

Clause 19.3 of JCT 1998 facilitates the selection of subcontractors using a methodology that affords the architect some choice in which subcontractor carries out the work, but without the need to use the nomination or naming procedures.

By using the clause 19.3 approach to selection, the contract documentation will include a selected list of a minimum of three subcontractors from which the contractor will make a selection using their normal subcontract tendering procedure. Once the contractor has made the selection the chosen subcontractor becomes a domestic subcontractor of the contractor. Other than the original selection of names, the architect has no further participation in the administration of the contract between contractor and subcontractor.

With domestic subcontracts, there is no mandatory requirement to use a standard form of subcontract, although Domestic Sub-Contract DOM/1 exists for that very purpose. Subcontract prices can therefore be obtained using the contractor's standard procedures and their own subcontract conditions as this basis of the enquiry documentation.

In choosing a domestic subcontractor the contractor will initially seek quotations at the time of carrying out the original estimate. During this tendering period the contractor will forward all necessary documentation to the subcontractor, including such items as original contract preliminaries which will indicate the general contract provisions, the specification relating to the subcontract works and the appropriate items from the bills of quantities. The package should also include the tender drawings necessary for the subcontractors to have a full picture of the job in hand.

This documentation should be sent to the subcontractors under cover of a tender enquiry letter, which should set out matters specific to the project, such as:

- the site location
- names of consultants and employer
- the main contract and subcontract conditions and any amendments to same, including the main contract appendix denoting specific project information
- details of whether a fixed or fluctuating price is required
- a schedule of daywork rates applicable to the project
- the return date for tenders
- a general description of the works
- particulars of site access
- the contract period for the main contract works
- the start date and duration of the subcontractor's works
- the discount level required by the contractor
- details of attendances and services required by the subcontractor from the contractor.

As with the original concept of producing bills of quantities, the principle of sending full information to the subcontractor is that all those tendering are working from the same level playing field. This will facilitate better and more accurate assessments by the subcontractor which should, in turn, create a better and more accurate final result for the contractor.

To safeguard the contractor's position the enquiry letter should record that the subcontractor must not under any circumstances alter or otherwise change the original documentation. This stipulation must relate not only to conditions of contract, but also to quantities and more particularly to specification. Many subcontractors will attempt to amend the original specification to suit their own products or those of their favoured suppliers, and on occasions this may result in cheaper or inferior products being included within the subcontractor's pricing structure.

To save on time and cost many contractors will, as an alternative to sending out such items as full preliminaries and full sets of drawings, invite subcontractors to view the appropriate documentation in the contractor's office. This gives the contractor and subcontractor the opportunity to discuss the project in detail with each other before submission of the formal pricing documentation.

To avoid doubt at a later stage, the contractor must keep accurate records of what information was given to subcontractors and what information they were shown, as well as a record of all verbal discussions that took place during this initial tendering process. Such accurate recording of this information may seem excessive, but these such records will facilitate speedy resolution of problems that may arise at a later date when final selection is being made.

Labour-only and Labour and Plant-only Subcontractors

The proposed use or selection of labour-only subcontractors will very much depend on the contractor's other commitments. If, for example, all of the contractor's operatives are engaged elsewhere, there may be a need to bolster the labour force through the use of labour-only operatives, particularly if labour availability in the area in question is limited.

The selection and administration of labour-only subcontractors is dealt with using the same approach as that of domestic subcontractors, but the contractor must be aware of the shifting nature of the labour-only market. As with other domestic subcontractors careful selection is paramount, not just in terms of pricing levels but also in areas of quality and the subcontractors' ability to perform to programme. With new subcontractors' references should always be sought and taken up, preferably from other reputable contractors working in a similar area.

The contractor should also obtain proof of the subcontractor's tax status, i.e. which Construction Industry Scheme (CIS) certificate they hold and the adequacy of their insurances. It is important to ensure that the following matters also form part of the subcontract package:

- What attendances will the subcontractor require, including plant requirements?
- Who is responsible for unloading material or, more importantly, the placing of such material at the point of use?
- Does the labour-only subcontractor provide supervision?
- Who provides and takes responsibility for the setting-out facility, if required?
- What is the frequency of payments to be?
- Does the subcontractor pay their own CITB contribution? If not, the contractor will have to make provision for collecting this.
- What retention fund will be retained by the contractor, and when will the appropriate reductions and releases be made?

Many of these matters will be dealt with during the original tendering process, but before subcontract orders are placed it is necessary for the contractor's quantity surveyor to re-evaluate the position of all subcontractors. Pre-tender comparisons will have been made, but many things change from the estimate to placing of orders, late quotations may be received or subcontractors pricing other contractors may submit their quotations on finding out who the successful tendering contractor was.

The following matters need to be considered.

Price

Price is a very large consideration, but should definitely not be the only criterion to be reviewed. However, it is necessary to compare all subcontractors who have submitted quotes with the rates included in the priced bills of quantities or work schedules. To do this it is necessary to complete an item-by-item check of all work sections involved, to ensure that all offers have been made on a like-for-like basis. When pricing, many subcontractors will return their tenders in a different format to the enquiry document sent out. This could be a lump sum for the job, page totals, or just items and rates scheduled out with no calculated extensions. It is therefore important that all items are checked. A simple computerised or manual spreadsheet can be used to complete the comparison (Figure 4.1).

This comparison will instantly highlight who has priced which item and whether anything requires further action.

Murribell Builders Ltd
Subcontract Comparison

| Project | City Centre Hotel | | Project Nr. | 12345 | | | | | |

Subcontract Type: Ceramic Tiling

Bill Reference	Quantity	Murribell Builders		Clark Ceramics		Gregory Wall Coverings		Blyth Tiling Co	
		Rate	Total	Rate	Total	Rate	Total	Rate	Total
Page 10/13 - A	1369	19.00	26011.00	18.50	25326.50	15.53	21260.57	18.50	25326.50
Page 10/13 - B	561	3.00	1683.00 incl		0.00	0.00	0.00	0.00	0.00
Page 10/13 - C	89	5.86	521.54	9.00	801.00	9.93	883.77	8.92	793.88
Page 10/13 - D	78	21.00	1638.00	17.00	1326.00	17.00	1326.00	14.67	1144.26
			0.00		0.00		0.00		0.00
			0.00		0.00		0.00		0.00
			0.00		0.00		0.00		0.00
			0.00		0.00		0.00		0.00
			0.00		0.00		0.00		0.00
			0.00		0.00		0.00		0.00
			0.00		0.00		0.00		0.00
			29853.54		27453.50		23470.34		27264.64
Adjustment for discount				2.5%	686.34 Net		0	1.75%	477.13
					26767.16		23470.34		26787.51
Lowest Net Subcontract			23470.34						
Residue/Shortfall			£6,383.20						

Figure 4.1 Subcontract comparision.

If a subcontractor has missed items that should be included within the subcontract package then the comparison will highlight this. Initially, the omission can be completed using either the bill of quantity rate for that item, an average of the subtrade competition or the highest of the competition, but by using at least one of the above the comparison can be made. Once the exercise is complete a judgement can be made on which subcontractors to review further. It is now necessary formally to fill in the gaps with those chosen subcontractors and discuss and agree the missing rates. When these have been formalised and discount has been taken into account, then further considerations can be made.

Tender Qualifications

The following statement should be included on the contractor's enquiry letter:

THE SUBCONTRACTOR SHOULD NOT ALTER OR OTHERWISE QUALIFY THE TEXT OF THE ENQUIRY DOCUMENT. ANY ALTERATIONS TO THE TEXT WILL BE IGNORED AND THE ORIGINAL TEXT WILL BE USED IN THE ANALYSING OF THE SUBCONTRACT PACKAGE AND THE PLACING OF ORDERS

Despite this statement, no matter how big and bold it appears, some subcontractors will always alter the enquiry document. If the same approach was made to the employer by the main contractor then their tender would be disqualified. It is therefore very important to check thoroughly the contents of each enquiry for subcontractor qualifications and amendments.

The most frequent alterations are *specification changes*, which can dramatically affect pricing levels. There will be many occasions when there will be a need to review specification, but in a straightforward tender situation the original specification must be adhered to. Many subcontractors and indeed many contractors will use such qualification to attempt to achieve an edge on their competitors, but this practice should be discouraged in the initial tender pricing scenario.

These reviews of the specification can be very useful when it comes to value engineering the project. A schedule of prospective alterations to the specification may be formulated for subsequent discussion with the architect and client's quantity surveyor, but this should be completed as a separate exercise.

It is also important to check for what attendances are required; these should be checked at tender settlement, but the subcontractor in prime position may have changed. All attendances required should be scheduled in a similar manner to the general pricing, i.e. checked against priced bill of quantities allowances. The list here is lengthy, but may include items such as:

- space for the subcontractor's site establishment and storage requirements, or general messing facility
- lighting, power and water supply
- scaffolding, both standing and special scaffolding requirements
- hardstandings for craneage
- hoisting
- unloading, storing and distribution of materials from stores to point of use
- placing of materials at the point of work for labour-only subcontractors

- fuel for testing equipment
- general building material, e.g. sand and cement for bedding roofworks
- heating and drying: many materials, e.g. ceiling tiles, require a warm environment
- clearing away and disposal of rubbish
- general builders' work, e.g. cutting holes and chases
- protection to completed works, e.g. floor coverings and glazing.

All such matters must be reviewed carefully and form part of the final analysis of cost for each subcontractor. Many of the attendances and qualifications made or required by subcontractors come with considerable cost to the project and the hierarchy chart may be altered when all such matters are considered. That aside, the contractor's quantity surveyor must be aware of the implications of such requirements so that provision can be made for them, regardless of whether or not allowance was made within the original tender documentation.

Conditions of Subcontract

Despite the fact that the contractor's original enquiry letters will state specifically the conditions of subcontract to be used and that the subcontractor must make due allowance for their inclusion within the subcontract package, subcontractors often include their own schedule of conditions. It is necessary, therefore, for the subcontractor's conditions of subcontract to be reviewed. Although most contractors have their own versions, any divergence from the original tendered requirements needs to be reviewed, discussed and fully agreed with the subcontractor before any order is placed.

Programming

It is vital to discuss with the subcontractors the anticipated programming of their works. Many employers demand almost instant starts to projects after acceptance of tenders. It is imperative, therefore, to discuss programme requirements fully and agree starting and completion dates, as well as duration periods if the subcontract is carried out in various sections, before any order is placed. There is no merit in having a subcontractor who can meet all the other criteria of price, quality, etc., if they are unable to meet programme requirements.

An important factor here is to analyse not only the job in hand, but also the subcontractor's commitments elsewhere. While the subcontractor may only have a small commitment to the contractor at that juncture, they may be fully committed elsewhere with other contractors. It is up to the contractor's team to seek out such problems, since subcontractors keen to forge new relationships, or indeed those with long ties with a contractor, will not want to lose any project. From both perspectives it is better that common sense prevails in such situations. Stretching a subcontractor past capacity will not help the project and may not only ruin relationships between the contractor and subcontractor, but also affect the contractor's association with the employer.

Particular note should be made of any special materials requirements, some of which may be on extended delivery. At this point there may be a need to consider materials other than those originally specified. JCT 1998 clause 8.1.1 requires that 'all material and goods shall, so far as procurable, be of the kinds and standards described in the Contract Bills'. This condition has regard not only to the general supply but also to the timely supply of materials. If specified materials are not available to suit the contractor's programme then the architect should consider the alternatives. All of these matters must be resolved before orders are placed and while an element of competition is still involved in the negotiations. When all of the above issues have been completed, the subcontractor's previous performance should be reviewed in terms of both accounting and performance on site:

- Did the subcontractor perform to programme?
- Was the subcontractor's quality of work up to the standard required? Did they keep the necessary quality records?
- What level of remedial work has been required on previous projects?
- Did the subcontractor comply with all safety requirements?
- What problems have been experienced in the production of interim valuations, variations and final accounting?

Once the above queries have been satisfied and full discussions have taken place between the contractor's and subcontractor's teams, a subcontract order can be placed. These discussions may be in the form of formal interviews, either during the tender period or during negotiations, or they may be conducted using a workshop scenario, where the construction teams can fully debate the way forward on the project as an integrated team.

Placing of Subcontract Order

A typical subcontract order is shown in Figure 4.2 and must include, apart from general project details:

- a full schedule of bill of quantities items included within the order value, including not only measured work sections but also full details of preliminary and preamble or specification pages;
- a schedule of all drawings sent to the subcontractor at tender stage or discussed with the subcontractor during the finalising of the subcontract negotiation;
- the date for commencement and completion of the subcontract works and the duration of these or sections thereof;
- the total subcontract price for the project, including any main contractor's discounts that have been negotiated;
- a statement as to whether the project is on a fixed price or a fluctuating price basis and, if fluctuating, the method of calculating this price;
- the conditions of subcontract applicable to the order and qualifications agreed between parties;
- the subcontractor's full and accurate details, i.e. name and address, for legal reasons;
- details of any correspondence that has taken place during precontract negotiations;
- project-specific details, such as retention and daywork percentages applicable to the project.

72　Commercial management in construction

Murribell Builders Ltd

Subcontract Copy
Order Nr
Date
Our ref

High Street
Sunniside
England
E1 2BB

Subcontractor

Subcontractor's Contacts
Manager
Telephone
Fax
Mobile
e-mail

Project

Contractor's Contacts
Manager
Telephone
Fax
Mobile
e-mail

PLEASE CARRY OUT THE _____ work as a DOMESTIC Subcontractor
all in accordance with the Main Contract Conditions, General Preliminaries
Section
Trade Preamble Sections
Bills of Quantities Pages
Drawing Nr.

at the rates contained within your quotation dated
reference
all for the sum of
less main contractors discount

Programme   Date for Commencement
             Date for Completion
             Duration

Subcontract Conditions   Murribell Builders Ltd
                          Standard Conditions

NB
This is a fixed price contract, increases or decreases in labour, material
and delivery charges will not be allowed

Signed

Quantity Surveyor   Contracts Manager

Figure 4.2　Subcontract order.

Once all of the above matters have been considered, checked and fully discussed with the contracts manager the order can be finalised. Both the contractor's quantity surveyor and the contracts manager should sign the authorisation for the subcontract order as confirmation that they are both in agreement regarding its contents. A further requirement is that the subcontractor returns a signed copy of the order to the contractor, acknowledging their agreement to the order and the terms and conditions contained therein. The contractor's quantity surveyor should also keep an updated record of orders sent and acknowledgements received to ensure that the necessary paperwork is kept up to date.

Subcontract Payments

Once the project is underway the duties of the contractor's quantity surveyor include monitoring, assessing and sanctioning payments to subcontractors.

Once interim valuation dates have been agreed with the client's quantity surveyor the contractor's quantity surveyor should circulate the appropriate dates to all subcontractors. When advising these dates the contractor's quantity surveyor should also request that the subcontractors should submit their applications for payment a day or two in advance of the agreed dates. As well as agreeing valuation dates at the start of the project it is useful to agree a format in which valuation should be submitted by the subcontractor.

Time spent wisely here will save valuable time once the project is underway. If, for example, stage payments are to be used as the valuation methodology within the main valuation, then it would generally be appropriate if a similar analysis were to be used with the subcontractor's valuation applications. This methodology will assist greatly in the preparation of cost value comparisons.

Even if the subcontractor does not submit an application, it is still the duty of the contractor's quantity surveyor to incorporate all subcontract matters in each interim valuation.

As mentioned previously, most main contractors will have their own subcontract conditions which will contain their own payment rules. However, many subcontractors, particularly the larger ones, will resist the inclusion of such terms with their order and a set of subcontract conditions including payment terms will have to be agreed. Indeed, many subcontractors request that the subcontract conditions should be those embodied in Domestic Subcontract DOM/1. Within the body of DOM/1 under clause 21, the payment terms are set out as follows.

Clause 21 of Domestic Subcontract DOM/1

21.1   The first and interim payments and the final payment shall be made to the subcontractor in accordance with clause 21.
21.2.1 The first payment shall be due not less than one month after the date of commencement of the subcontract work on site.
21.2.2 Interim payment shall be due at intervals not exceeding one month calculated from the date when the first payment was due.
21.2.3 The first and interim payments shall be made not later than 17 days after the date when they become due.

It is important, therefore, to keep an eye on the timing of payments to all subcontractors, particularly those whose subcontract conditions include the above

Figure 4.3 Subcontract payment.

CONTRACTOR	MURRIBELL BUILDERS LTD HIGH STREET SUNNISIDE ENGLAND E1 2BB

SUBCONTRACTOR	

CIS CERTIFICATE DETAILS

DATE	
TYPE	
NUMBER	
EXPIRY	
HOLDER	

VAT REG
CONTRACT
CONTRACT Nr.

	Certified	Current Payment	Previous Payment	Nett Payment
Gross Payment				
Retention %				
Discount %				
Net Payment				
VAT %				
TOTAL DUE				

THIS IS YOUR OUTPUT TAX FOR WHICH YOU ARE ACCOUNTABLE TO CUSTOMS AND EXCISE

Figure 4.3 Subcontract payment.

provisions. The basis of the subcontract payment is the '*subcontract liability,*' which should be calculated as described in Chapter 5.

The payment document (Figure 4.3) will generally be computerised and all the contractor's quantity surveyor needs to do is to formulate a subcontract payment request schedule (Figure 4.4) indicating the various figures to be paid, which in gross terms should equate to the gross subcontract value indicated on the subcontract liability schedule (see Figure 5.3). The schedule should also include the subcontract order value and the discount to be taken from the payment, both figures coming directly from the subcontract order. The recording of the subcontract order value creates a valuable tool to be used as a monitor to the contractor's quantity surveyor and also to those who are countersigning the payment,

Subcontractors 75

Murribell Builders Ltd

SUBCONTRACT PAYMENT REQUEST

Contract

Nr.

Date

Certificate Nr.

Payment Sheet Nr.

Amount Due

Subcontractor	Type	Order Value	Amount of Subcontract Claim	Gross to be Paid	Gross Previously Paid	Ret %	Disc %	VAT %	Net Amount Paid	Date Paid	Remarks

Signed
Quantity Surveyor   Date   Authorised   Date

Figure 4.4 Subcontract payment schedule.

generally the contracts manager, so that a quick check can be made to see that payments are not exceeding the original order value.

If payment figures do exceed the original order value, then the contractor's team can discuss why that should have occurred. Have extra works been carried out on site with or without proper instruction; or has remedial work been instructed at site level to overcome an operational problem? Whatever the reason, the contractor's team will be alerted to investigate why extra payment is being sanctioned.

If it is found that the subcontractor is claiming for additional works where no instruction has been received, then further action will need to be taken by either the contractor's quantity surveyor or the contracts manager in obtaining the necessary architect's instruction.

The subcontract payment request schedule should also indicate the retention percentage to be deducted from the payment, details of which were sent to the subcontractor with the original enquiry. This figure should also reflect the same percentage as the main contract retention fund, including any reduction in the fund amount resulting from the employer taking partial possession of a section of the works or at practical completion.

Construction Industry Scheme (CIS)

The new CIS started on 1 August 1999, replacing the old 714 scheme and rendering void certificates issued on the old scheme. The new scheme was introduced by the Inland Revenue to facilitate a more controlled tax collection administration within the industry. The new scheme requires considerable input on the part of both contractors and subcontractors, with the issuing of tax certificates being dependent on more accurate and in-depth information being supplied by the applicant. However, to obtain a registration card, as a prerequisite of payment, the subcontractor needs only to complete a form, supply a photograph and attend an identity check.

Under the scheme a contractor has to make a deduction in certain situations when making a payment to a subcontractor under a contract relating to construction operations.

There are two types of registration card.

- The permanent registration card CIS4(P) bears a National Insurance (NI) number or, exceptionally, a 'system identifier', which is a number created by the Inland Revenue for those individuals who do not need an NI number, such as non-residents. The CIS4(P) does not have an expiry date.
- The temporary registration card CIS4(T) does not bear an NI number and has an expiry date.

There are also two types of certificates.

- The subcontractor's tax certificate CIS6 is issued to a qualifying individual in business on his or her own, a qualifying partner in a firm, and directors and company secretaries of companies entitled to gross payment.

- The construction tax certificate CIS5 is issued only to companies that meet certain requirements.

Businesses in the construction industry are known as contractors or subcontractors. These businesses can be companies, partnerships or self-employed individuals. Both contractors and subcontractors must hold a valid registration card or subcontractor's tax certificate before they can be paid under the new scheme. To obtain a registration card or tax certificate the contractor or subcontractor must be registered with the Inland Revenue. Subcontractors who meet the qualifying conditions will be issued with subcontractor's tax certificates by the Inland Revenue, while those who do not will be issued with a registration card. Where the subcontractor holds a registration card, the contractor must deduct from all payments for labour an amount on account of the subcontractor's tax and national insurance contribution (NIC) liability. Where the subcontractor holds a tax certificate, payments made by the contractor should be made gross.

All contractors and subcontractors within the construction industry are affected by the scheme. As far as the scheme is concerned 'contractors' means not only contractors but also government departments, local authorities and clients. Private householders and businesses that spend less than £1 million a year on construction work are *not* classed as contractors and are therefore not included in the scheme. Subcontractors are defined as businesses that carry out building work for contractors. The scheme also facilitates the situation where companies act as both contractor and subcontractor. This means that they both pay businesses below them and are paid by businesses above them in the chain. The scheme covers all construction work and also jobs such as installation, repairs, decorating and demolition.

It remains the contractor's responsibility to establish whether the person is employed or self-employed. If he or she is employed, the normal pay as you earn (PAYE) system must be operated. If not, then it is the contractor's responsibility to ensure that the subcontractor's registration card is genuine and that the person producing it is the user of the card. Equally, the contractor should establish whether the registration card is a permanent or temporary one. If the card is a temporary registration card CIS4(T) the contractor must make a record of the expiry date, after which no payments must be made to the subcontractor, unless he or she has since obtained and can show another valid registration card or subcontractor tax certificate.

5 Cost Value Comparisons

Cost value comparisons, or reconciliations, are usually completed by the contractor's quantity surveyor, but the process will require liaison with other departments in its completion. In addition, considerable discussion is required with the rest of the project team, i.e. the contracts manager and the site manager, before the cost value comparison is issued to the rest of the management team.

There are two basic reasons for conducting cost value comparisons in contracting organisations: one is good management practice and the other is accountancy requirements.

While there are no legal requirements to enforce companies to monitor their performance using a set methodology, the Institute of Chartered Accountants in England and Wales has issued guidelines which direct companies to follow certain parameters. This document, the 'Statement of Standard Accounting Practice No. 9 (SSAP9)' comprises a series of explanatory notes that are intended to remove inconsistencies from financial reporting procedures relating to published or financial accounts.

To satisfy statutory rules companies are required to have their financial accounts audited on a yearly basis, and companies run the risk of having their accounts qualified by their auditors if they cannot demonstrate that they have complied with the requirements and guidelines of SSAP9.

It is essential therefore to any organisation that management are kept informed on a regular and frequent basis of how things are going in terms of company performance, whether this is profitable or not. In simple terms, within contracting organisations, this amounts to comparing the actual costs to be allocated to projects against accurate assessments of the value of works carried out, measured and valued at contract rates, and the value of materials and goods properly delivered and stored on site.

The SSAP9 guidelines advocate a conservative approach to completing annual accounts and cost value comparisons. They divide their considerations into two sections: long-term projects, i.e. contracts that span more than one accounting year, and short-term projects, i.e. those that do not.

If, for example, costs are incurred on a project before the end of an accounting period but no valuation is carried out and therefore no income is available, then, provided reasonable evidence can be shown that the costs will be covered by future revenue, it is acceptable to carry the costs forward into another costing period rather than anticipating revenue to set against the costs.

The reasoning in suggesting such methodology may be that the costs or expenditure are provable and checkable figures, whereas anticipating income involves making subjective assessments, which may not be verifiable at the time of comparing figures or auditing accounts.

The results of these specific project costings can be used in many ways, either collectively or as a monitor of individual project performance. The collective results form part of the contractor's management accounting process. The management accounts are the contractor's method of reviewing their performance throughout any financial year and should not be confused with their financial accounts, which are a statutory requirement for any limited company. Among other things the cost value comparison can be used:

- to compare project profitability and turnover against budget and forecast figures (see Chapter 2)
- to monitor project performance in terms of labour, plant and material costs against original tender figures
- as a monitor of general performance to be used when assessing tenders on other similar projects.

To satisfy the requirements of SSAP9, and also those of the shareholders, banks and other funders, the contractor will be required to demonstrate performance, using sound management techniques, by setting budgets and monitoring actual performance against these targets. As indicated in Chapter 2, at the onset of each contract, the contractor's team should analyse the project in respect of anticipated net cost of production (NCP) and overhead and profit expectations, i.e. income and expenditure analysed monthly throughout the currency of the project. These individual assessments are drawn together to produce an overall picture of the company's projected performance.

Once projects commence, on-site project costings or cost value comparisons should be carried out on a monthly basis. The results of each project can then be monitored against the above anticipated performance. This comparison may be either against the company budgets set for the particular financial year, or against forecasts or revisions to the budgets, or both. In management accounting terms, however, cost value comparisons must be monitored against the original yearly budgets. The forecast, as described elsewhere, is the re-examination and updating on a regular basis, generally quarterly, of the original budget assessments.

To be of most benefit to the management team it is necessary for cost value comparisons to be produced as expeditiously as possible after interim valuations are completed and agreed on site. There is little point in producing cost information several weeks after the event, a situation which could render remedial measures ineffective.

It is pertinent at this stage to mention that cost value comparisons are not just completed to monitor performance; they are an essential management tool used in many departments, in particular estimating. Feedback from project performance is required continuously to ensure that accurate and, above all, competitive estimates can be produced. This situation would apply equally to a bona fide tender and to an initial brief cost plan.

The cost value comparisons can take many formats, but essentially, when comparing cost and value it is important to adhere to certain key parameters.

Both cost and value must be taken to a specific date. Many companies have set cut-off dates for costing, e.g. the end of the month, whereas valuations may, owing to the date for commencement of a project, fall at the midpoint of a month. It may therefore be necessary to adjust one or the other, or both sets of figures before a comparison can be drawn.

It is also essential to avoid overcomplex systems and reporting procedures that in many cases serve no useful purpose and can tie up valuable time for the project team. It is equally a waste of effort to complete these cost exercises if they are not formally discussed and acted upon by others. The cost reporting should be tailored to suit the management team's requirements, but the cost value comparison should also follow the principles and guidelines set out in SSAP9.

It is necessary to ensure that there is an accountability within the management team during the discussions relating to the cost value comparison. During these discussions, which to be effective must be regular, say at a monthly management meeting, all costs can be debated and action plans formulated and minuted. The establishment of the cost value comparison format should therefore be devised strictly on a basis of the need for essential information.

A specimen cost value comparison is shown in Figure 5.1. In brief, this example can be divided into four main sections. The top of the schedule contains general contract data, such as interim valuation and cost valuation comparison number and date, personnel involved with the project and contract dates, with the current situation recorded, the original tender sum and projected final account figure highlighted. The right-hand section deals with the cumulative value and cost figures including subcontractors, ending up with the gross profit or loss to date. The left-hand section represents the period changes in cost and profitability, with the middle section setting out original figures from the tender in respect of all the individual components of the estimate.

As can be seen from Figure 5.1, the cost value comparison contains considerable information, and its completion looks like a formidable task. However, many of the initial details on contract and tender allowance will only need to be sourced once for the completion of cost value comparison no. 1, and thereafter transferred from cost value comparison to cost value comparison. Valuation and cost figures will have to be reviewed and updated with each issue.

The cost value comparison schedule needs to contain elements that facilitate checking of other sections of the comparison, and should also contain information for use by other departments, such as accounts, general management and estimating.

To highlight these areas, let us look in more detail at the relevant sections of the schedule. The first line of the cost value comparison simply records the cost value comparison number and, importantly, the cut-off date up to which the cost has been assessed. This information allows a simple check to be made against the valuation date recorded in the next section.

The next section is interim valuation information.

>Certificate No
>Gross Value
>Valuation Date
>Retention

This section records, from the *architect's interim valuation certificate*, the valuation number, the gross amount of the interim valuation, the date of the interim valuation and the amount of the retention deducted from the valuation figure. A copy of the architect's interim valuation certificate should be held in the contractor's quantity surveyor's files in support of this section. This certificate will verify the retention figure and, just as importantly, verify the date of the interim valuation,

Cost value comparisons 81

	Murribell Builders Ltd	
	Cost Value Comparison	
Profit Appraisal No	at	

Certificate No	Contract	C.Manager
Gross Value	Location	Q.Surveyor
Valuation Date	Contract No	Agent
Retention	Estimator	Bonus

		Contract	Prelims
TENDER SUM	Contract Commencement Date		
Less Dir.pymts.	Contract Completion Date		
P.C.Adjustment	Contract Weeks Ahead/(behind)		
Variations	Revised Completion Date		

		Contract	Prelims
Projected F/Acc............£	Ext.App.For	Weeks	
Less Adjusted Value	Ext.Granted	Weeks 0.00%	0.00%
Outstanding Value........£	Ant.F/Awd	Weeks	
% Complete			

GROSS CERT VALUE
Sundry Invoices   0

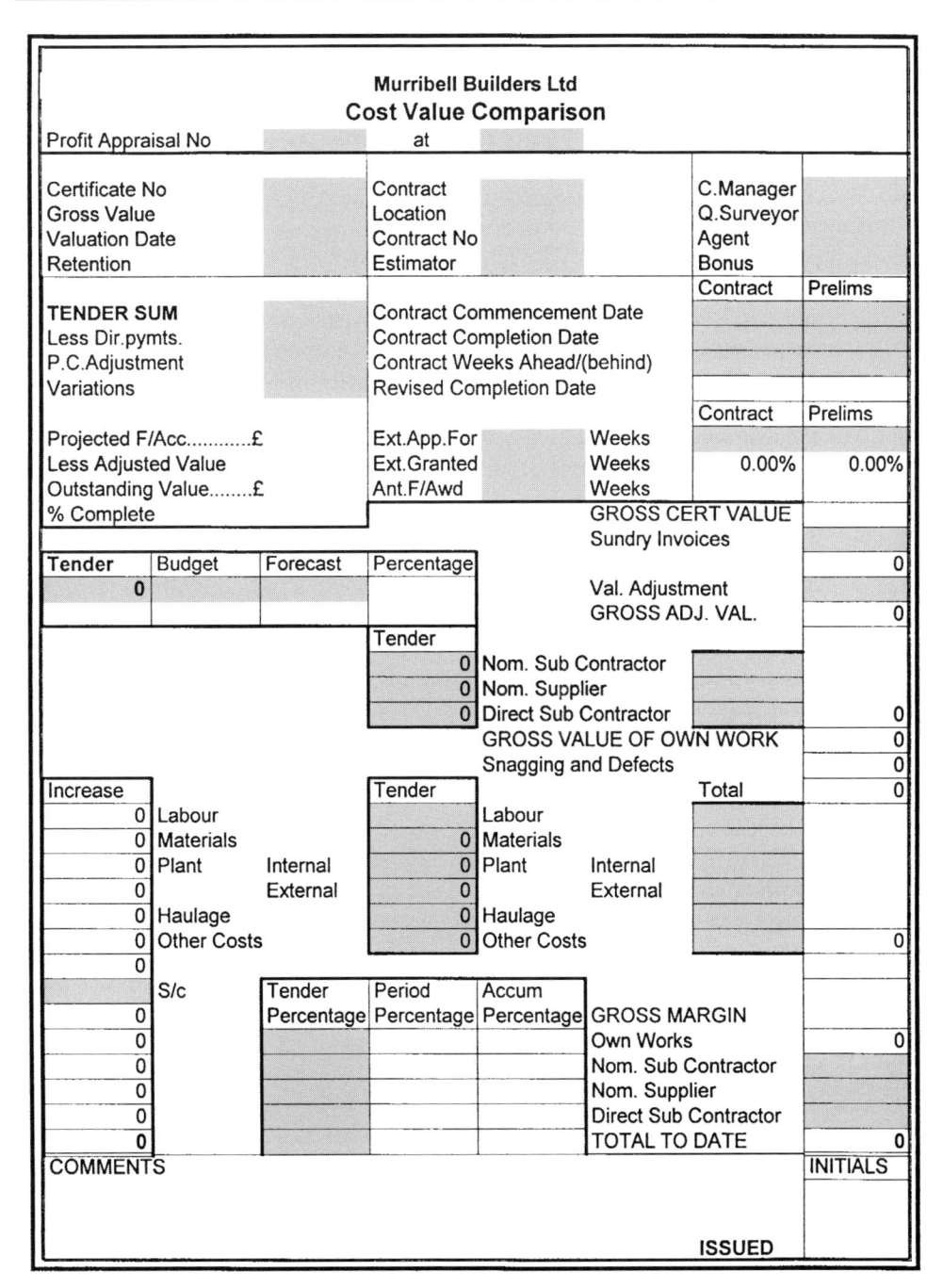

Tender	Budget	Forecast	Percentage
0			

Val. Adjustment
GROSS ADJ. VAL.   0

	Tender		
	0	Nom. Sub Contractor	
	0	Nom. Supplier	
	0	Direct Sub Contractor	0

GROSS VALUE OF OWN WORK   0
Snagging and Defects   0

Increase			Tender			Total	
	0	Labour		0	Labour		0
	0	Materials		0	Materials		
	0	Plant	Internal	0	Plant	Internal	
	0		External	0		External	
	0	Haulage		0	Haulage		
	0	Other Costs		0	Other Costs		0
	0						

	S/c	Tender	Period	Accum		
0		Percentage	Percentage	Percentage	GROSS MARGIN	
0					Own Works	0
0					Nom. Sub Contractor	
0					Nom. Supplier	
0					Direct Sub Contractor	
0					TOTAL TO DATE	0

COMMENTS		INITIALS

ISSUED

Figure 5.1 Cost value comparison.

which in turn should correspond to the date of the cost value comparison at the top of the schedule. If not, then a note should be made in the valuation adjustment section or the comments column of the cost value comparison explaining why any such variance has occurred.

The *gross value* from this section must always equate to the *gross certified value* figure to be inserted at the head of the right-hand column, which is in effect the beginning of the cost value comparison. To avoid errors, if the cost value comparison is computer generated, then the *gross certified value* should preferably be inserted as an automatic transfer from the valuation information section.

The recording of the retention figure in the cost value comparison facilitates the monitoring of the amount of the retention fund against the completion dates, which are also recorded in the cost valuation comparison schedule. As the completion date approaches, the contractor's quantity surveyor can discuss with the project team when the certificate of practical completion is to be issued, which should ensure the timely application for the release of monies from the retention fund.

Further contract information should be included as required. On the example used in this appraisal, contract personnel are recorded alongside the contract name and location details. Other specific contract detail is also registered; contract timing is recorded in terms of actual dates and also in weeks, with a percentage calculation expressing progress over the original contract period. The example shown allows for the reappraisal of the contract completion dates as circumstances warrant and also for information to be recorded in relation to any extensions of time that may be required.

If delays are registered in this section, then a corresponding allowance should be made in the valuation adjustment section to cater to this change in contract circumstances. In addition, if nothing is recorded in the section relating to extensions of time and yet a delay is recorded elsewhere, management can review the situation. By assessing the principles of the delays, they can consider whether appropriate notices should be sent to the architect recording the delay, or whether allowances have been or need to be made for overrun costs, and perhaps the levying of associated liquidated and ascertained damages should the delays be due to default on the part of the contractor.

The example shown in Figure 5.1 differentiates between actual contract dates and timings and the dates on which the project has been assessed by the contractor. Many contractors will re-examine contract periods when compiling their estimates, resulting in a different period to that recorded in the contract documentation being used in the contractor's estimate. The assessments made against both of these allowances included on the cost value comparison need, therefore, to run in tandem, assessing each on its own merit and taking the appropriate action as each set of figures requires.

So that a comparison can be made in respect of the total valuation expenditure when the cost value comparison is completed, and the contract programme timing, the example given in Figure 5.1 indicates an assessment of the projected final account figure. This figure is assessed using the original tender sum and updated in line with all known facts at the time of the review. Allowances are made for direct payments and the effect of the valuation of architect's instructions and the adjustment of provisional and prime cost (PC) sums.

By expressing the *gross adjusted valuation* figure as a percentage of this projected final account figure, a comparison can be made with the percentage complete figures highlighted in the contract date sections. While this assessment will vary greatly from job to job, it serves as a worthwhile point of discussion for the management team. On repeat housing projects, for example, there should be a parity between figures, whereas on one-off project such as schools, factories and offices, etc., the figures may be distorted, for example by the large values of mechanical and electrical installations that may be installed in a short space of time near the end of the contract period. Nonetheless, the comparison should be used particularly if the spending rate is considerably behind that of the programme timing.

Gross Certified Value

Moving to the cost valuation comparison itself, the starting point must always be the *gross certified value*. As already indicated, this figure must be supported by the appropriate architect's interim certificate. This is the external valuation and not the contractor's quantity surveyor's assessment or internal valuation of works carried out on site.

Valuation Adjustment

Below the gross certified value, continuing down the right-hand column of the cost value comparison schedule, it is generally necessary to adjust the external valuation.

The first adjustment to be made is that for *sundry invoices*, i.e. works carried out on the project which do not form part of the contract works. These may be direct works to the employer or works carried out for a subcontractor as a separate issue, but are works that will, in costing terms, be charged to the project. A copy of all invoices with an up-to-date summary should be included in the valuation/cost value comparison file.

The next item is the *valuation adjustment*. This reconciliation can be wide and varied, and must include all adjustments, either undervaluations or overvaluations, necessary to bring the external interim valuation to an accurate *gross adjusted valuation*, which can then be used for costing purposes. The development of this schedule needs to be thorough and well documented, not only in terms of producing the required level of accuracy, but also to facilitate ease of checking by others should this be necessary.

An external valuation may require adjustment for many reasons, not least arithmetical errors found in the external valuation after agreement with the client's quantity surveyor. Common areas of adjustment are as follows.

1. Adjustments for external preliminaries claimed in valuations against the internal preliminary schedule previously described in the interim valuation section (see Chapter 3).
2. Adjustments for elements included within costing but not in the external valuation. A prime example is materials brought to site on the same day as the valuation, but after the materials on site were recorded, but nonetheless included with the project costings to the cost valuation date. In this instance, while it is customary to add the value of such materials to the valuation adjustment, if working strictly in accordance with SSAP9 the value of the material should be extracted from the cost section of the cost value comparison.
3. Simple overzealous inputting into the external valuation of works by the contractor's quantity surveyor and not picked up by the client's quantity surveyor. Items of overmeasurement or elements of work claimed within

interim valuations must be invalidated within the valuation adjustment figure for reintroduction into the revenue side of the cost value comparison once the works have been fully completed on site.
4. Any adjustments necessary to bring the cost cut-off date and on-site valuation date together. It is essential that both interim valuation dates and the date of assessing costs correspond. When making such adjustments, although it may sometimes be more convenient to attempt to reassess the interim valuation figure, a more accurate assessment will be obtained if the cost figure is reviewed and adjusted accordingly.
5. When dealing with variations, even if the valuations and architect's instruction schedules are up to date, in many projects there will be areas of the schedule that have not been fully agreed with the client's quantity surveyor. In some instances their valuation will differ considerably from that of the contractor's quantity surveyor. SSAP9 requires prudence in these instances. In addition, as projects are completed, architect's instructions will be given without formal written recording taking place. This situation will occur, no matter what the conditions of contract requires. The clerk of works may issue instructions that are yet to be confirmed or that the architect may issue verbal directions covering extras to the contract, again without confirmation. The contractor's quantity surveyor may record these in the interim valuation and the client's quantity surveyor may include an allowance within the valuation, against the contract rules. In such instances the contractor's quantity surveyor must err well on the side of conservatism when using such assessments in the cost value comparison. The prompt here is that the contractor's team should seek immediate issue of any outstanding architect's instructions.
6. Contractual claims for loss and/or expense that have not been agreed with the client's consultants, no matter how strong the contractor thinks his claim is, should be excluded from any revenue figures.
7. The reverse of the above is the situation where a project is in a delay situation and will run past its date for completion, without extensions of time being agreed with the architect and extension of time certificates being issued. Provision should then be made within the valuation adjustment to cover the eventuality of liquidated and ascertained damages being claimed by the employer.

   Again, as with item 5, the prompt is for the contractor's team to conclude matters referred to in items 6 and 7 as quickly as possible.
8. Provision should also be made in the valuation adjustment for future known losses. If, for example, initial works on a project are profitable but known errors exist in the pricing of works to be carried out at a later date, then provision should be made in the valuation adjustment to cater for this occurrence.

The adjustment of the external valuation is paramount to the end result but must not be taken as *carte blanche* to add spurious items of claim, whether in pricing of variations or additions made because 'we thought that figure should be higher'. Accuracy *must* be the only policy. Any other figures will be highlighted, if not in the next costing, then certainly before the final account is complete. Imaginative accounting will always be a short-term, and absolutely a no-win answer to profitable costs.

The general principle for the contractor's quantity surveyor to remember must be that of caution. Any figures included in the valuation adjustment must be capable of substantiation and should, wherever possible, have been agreed with the client's representatives before inclusion in the valuation adjustment figures.

Deductions

Once the *gross adjusted valuation* has been assessed three further elements are deducted from the figure to arrive at a final residual value or margin (profit).

Subcontract Liabilities

The first of these is the *subcontract liabilities* for all disciplines of subcontractor, i.e. nominated or domestic/direct subcontractors and nominated suppliers, but excluding labour-only subcontractors, the cost of which should be included later in the labour section of the main contractor's costing.

The subcontract liability is a schedule of works per subcontractor included within the external valuation, calculated at the rates contained within the contractor's bills of quantities and also at the subcontractor's rates, effectively drawing a comparison between what the contractor has been paid and what the contractor is liable to pay to any given subcontractor.

This liability should include not only all those matters listed in the external valuation, such as contract works, variations and materials on site, but also any works that the main contractor is due to pay the subcontractor but for which he would not receive reimbursement for through the main works contract provisions (Figure 5.2).

If any valuation adjustments have been made to the external valuation appertaining to subcontract works, then provision will equally have to be made to the subcontract liability section of the cost value comparison.

As with the main valuation adjustment, the contractor's quantity surveyor should be cautious in the assessment of the subcontract liabilities. Situations may arise when interpretation by the client's quantity surveyor of the valuation of an architect's instruction may differ from that made by the contractor and from the subcontractor's assessment. As indicated in Chapter 3, the valuation of variations rests solely with the client's quantity surveyor, although the valuation rules contained within clause 13 of JCT 1998 must be followed. When assessing the valuation of subcontract variations, therefore, the figures of the client's quantity surveyor should be the starting point.

The results from the individual subcontract liabilities in Figure 5.2 can be summarised on the schedule shown in Figure 5.3.

Snagging and Defects

The next deduction to be made should be that for *snagging and defects*, and this can be subdivided into two sections:

- snagging required at the end of the project to achieve handover
- a levy to be used to cover for costs that may be incurred in the making good of defects period before the certificate of making good of defects is issued.

Murribell Builders Ltd

Subcontract Liability

Contract		City Centre Hotel		Valuation Nr.	2
				Date	12/12/97

Excavation & Earthworks

Page	Item	Bill Quantity	Valuation Quantity	Murribell Builders Ltd		Wright Excavations	
1/8	1	1333	1333	0.23	306.59	0.16	213.28
	2	400	400	0.70	280.00	0.60	240.00
	3	400	200	10.00	2000.00	7.00	1400.00
19/2	1	20	16	2.00	32.00	2.50	40.00
	2	679	543	0.70	380.10	0.60	325.80
	3	679	543	10.00	5430.00	7.00	3801.00
	4	1579	0	0.15	0.00	0.15	0.00
	5	1579	0	4.00	0.00	4.05	0.00
19/5	1	453	362	0.70	253.40	0.60	217.20
	2	453	362	10.00	3620.00	7.00	2534.00
	3	1332	0	0.15	0.00	0.15	0.00
	4	1332	0	4.00	0.00	4.05	0.00
19/15	1	594	476	0.70	333.20	0.60	285.60
	2	297	238	10.00	2380.00	7.00	1666.00
	3	297	238	0.90	214.20	1.00	238.00
	4	297	0	1.50	0.00	0.80	0.00
1/1	1	40	40	30.00	1200.00	0.00	0.00
	2	20	20	20.00	400.00	0.00	0.00
	3	1	1	2000.00	2000.00	4780.00	4780.00
1/16	4	1	1	3852.00	3852.00	3852.00	3852.00
					22681.49		**19592.88**

NB Whilst every effort should be made to sublet works within Bill of Quantity allowances as page 1/1 item 3 demonstrates that is not always possible. It is important therefore to assess the subcontract as a whole bearing in mind the effect future variations may have on the project.

Figure 5.2 Subcontract liability.

Cost value comparisons 87

Figure 5.3 Subcontract liability summary sheet.

Both of these figures are highly dependent on the type of project involved and each should be judged on its own merits and by a comparison of previous records in these matters on similar projects. On housing or multiple unit projects, for example, an allowance per unit may be adopted, which would facilitate easy checking of the assessments as works are completed and handed over. Many contractors will have standard allowances for such elements, perhaps 1.5% of the contractor's gross value (excluding subcontract figures), building up over the contract period, then reducing to 0.5% after practical completion is achieved, as an allowance retained for making good works during and at the end of the period of making good the defects. These allowances should, however, be reviewed in joint discussion with the project team as each cost value comparison is considered. This consideration is of particular importance on larger projects where the use of a set percentage may result in higher than required figures being retained in the cost value comparison for snagging works.

The above adjustments are made only on the main contractor's work, as it is assumed that subcontractors will attend to their own snagging and defects work at their own cost.

Contractor's Core Costs

The third and final element of deduction is that of the *main contractor's core costs*, i.e. labour, material and plant, and other associated costs. In general, these will be supplied to the contractor's quantity surveyor by a separate cost department within the contractor's company. By adopting this method of independent cost assessment, the contractor's quantity surveyor is spared the arduous task of completing such costs, but also, and more importantly, the management team will have the comfort that the costs have been produced by an independent section outside the immediate project team, with no specific interest in the resultant outcome of the cost value comparison. It should, however, remain the duty of the contractor's quantity surveyor to analyse these costs during the completion of the cost value comparison.

The specimen cost value comparison indicates six levels of cost:

- labour
- material
- plant (internal)
- plant (external)
- haulage
- other costs.

However, this schedule can be as long or as short as needs dictate.

Labour and materials, for example, can readily be subdivided into various elements:

- excavation
- drainage
- brickwork
- joinery work
- concreting

depending to a large degree on how sophisticated the contractor's costing system is and what level of information is required by the management team to complete their function.

The example shown subdivides the costs only into basic elements, but can be viewed against the original tendered targets highlighted in the centre of the cost value comparison and, when used in conjunction with 'percentage complete' figures from the *tender sum analysis* box, a view can be taken on how various sections are performing as the works proceed.

If no costing department is available then the contractor's quantity surveyor must produce costs manually using material and plant returns from site and comparing these with invoices received and orders placed, and scheduling labour and staff charges, including all costs such as National Insurance and holiday pay.

The importance of the accurate and timely completion of the plant and material returns by the project team cannot be overemphasised. These returns should record in detail all plant and materials delivered to site. It is not good practice when completing these returns to group items together; a little extra time spent in completing these documents properly will save others considerable effort in completing their costing function. The returns should be completed on a weekly basis and should be passed to the costing department for their further action.

The final section of core contractor costs is that of 'other costs'. This section is not just a balancing section. Elements such as fees and payments to statutory authorities should be scheduled and included in this section, as can payments to labour-only subcontractors or scaffolding costs, should it be felt necessary to split these figures from the other main cost sections to make future analysis easier.

Once all these figures are known and inserted into the cost value comparison the residue entered into the own works section of the *gross margin* is the profit generated from the main contractor's work element of the project, and this, when added to the profits made in other sections, i.e. the subcontract liability schedule, indicates the total profit to date from the project. This figure can then be compared with the original contract profit included in the contractor's tender, and with the further assessments included in the contractor's yearly budget and the current forecast for the project. These figures appear below the assessment of the projected final account section of the cost value comparison schedule.

To complete the analysis of the schedule, the central band is a record of the original tender allowances from which the management team can compare current cost levels. The figures to the lower left of the schedule are used to indicate the period movement in costs from the previous cost value comparison. The lower central area conveys the various results in percentage terms for ease of comparison against the recorded original tender figures.

The last boxes are the comment and signature sections. Once completed, the cost value comparison should be discussed at length with the contracts manager and site manager, and where necessary an appropriate action plan formulated. The signatures are required as confirmation that the cost value comparison has been discussed and that the project team is fully aware of the current situation on site.

When these discussions have taken place the necessary comments should be inserted in the cost value comparison schedule. Once the schedule has been signed by all parties as being agreed, checked as complete, it can be circulated to the remainder of the management team.

As indicated previously, the completed cost value comparison schedules can then be used to form the basis of the contractor's management accounts. These accounts, generally completed monthly, gather all cost value comparisons together to produce an overall perspective of profitability. The collective totals produce the gross profit from which the contractor's assessment of monthly overheads is deducted, the resultant figure being the contractor's assessment of profitability (see Figure 2.5 in Chapter 2).

For these collective assessments it is customary that they are produced with totals for:

- the month currently being reviewed
- the year to date
- the total project performance to date.

These figures can then be compared with the contractor's original budget figures and against any reforecasting that may have been carried out.

As described at the start of this chapter, the timing relating to the production of cost value comparisons is crucial to the management team. Once an interim valuation has been completed and agreed with the client's quantity surveyor, preparation work should begin. In practical terms, the contractor's quantity surveyor should try to complete the cost value comparison within two weeks of the valuation agreement date. They should ensure that the costing department has the information required from the project on-site team so that they can comply with the timing requirements for completion of the cost value comparison.

As with information produced for budget and forecast revisions (see Chapter 2), simple graphs should be used to highlight the results. These graphs will clearly show any developing trends and can be used as discussion points during project team meetings (Figure 5.4).

The examples shown in Figure 5.4 compare various elements but should not be regarded as finite; they are simply suggestions of some areas that could be identified to assist the project teams in the running of each project.

From these examples the following inferences can be drawn.

- Top left   This chart analyses forecast sales or total valuation against actual figures, and also considers anticipated net cost of production against actual cost to date. The chart in this instance identifies that both spend and cost turnover are running ahead of schedule. If the levels of value and cost were not in the same ratio, particularly if cost were being expended at a greater level, then the project team would have been alerted to the problem and would be able to formulate the necessary action plan to evaluate that particular situation.
- Top right   This chart analyses anticipated overhead and profit expectations against actual figures. As can be seen from the chart, all is well, with profit being returned at a higher rate than anticipated.

Cost value comparisons 91

City Community Centre

month	forecast sales	actual sales	forecast ncp	actual ncp	month	oh/p	actual	month	labour - est.	labour - antic.	labour - actual	month	plant - est.	plant - antic	plant - actual
Jan	0	151102	0	145562	Jan	0	5685	Jan	9847	1600	3064	Jan	0	2586	4026
Feb	170000	289672	165000	267177	Feb	5000	21115	Feb	38075	17200	20203	Feb		7887	12822
Mar	435000	694086	415000	661531	Mar	20000	32555	Mar	87722	73600	41186	Mar		15334	24847
Apr	730000		690000		Apr	40000		Apr	120930	107600		Apr		24358	
May	1057000		990000		May	67000		May	166004	156000		May		34388	
Jun	1397000		1300000		Jun	97000		Jun	191042	178400		Jun		44856	
Jul	1727000		1600000		Jul	127000		Jul	206940	191200		Jul		55191	
Aug	2028000		1875000		Aug	153000		Aug	225344	204400		Aug		64824	
Sep	2304000		2125000		Sep	179000		Sep	233194	209200		Sep		73186	
Oct	2589000		2375000		Oct	214000		Oct	238434	210800		Oct		79706	
Nov	2855000		2605000		Nov	250000		Nov	243674	212400		Nov		83816	

Programme		
CVC Nr 1		- weeks
CVC Nr 2		- weeks
CVC Nr 3		- weeks
CVC Nr 4		
CVC Nr 5		
CVC Nr 6		
CVC Nr 7		
CVC Nr 8		
CVC Nr 9		
CVC Nr 10		
CVC Nr 11		

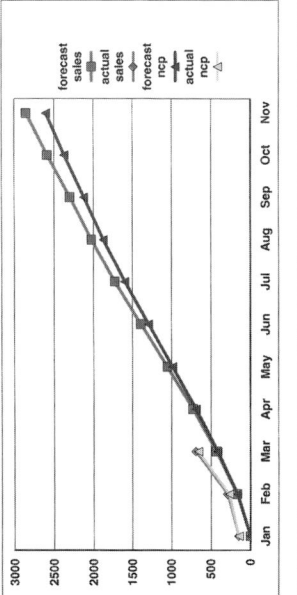

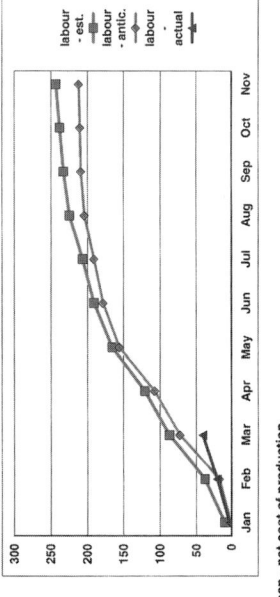

ncp - net cost of production
oh/p - overheads and profit

Figure 5.4 Comparsion graphs.

■ Bottom left and right	This trend is following the increase in turnover identified on the previous graph. These charts identify and analyse actual expenditure on both labour and plant against anticipated figures from the original estimate. The labour element goes a stage further by also comparing actual and original estimate figures against the project manager's assessment of labour requirements. In this example, while the labour figures suggest that all is well, the plant analysis reflects an overspend and will certainly require investigation.

Recommended Further Reading

Barrett, F.R. (1981) *Cost Value Reconciliation*. Chartered Institute of Building.

Cottrell, G.P. (1978) *The Builder's Quantity Surveyor*. CIOB Surveying Information Service No. 1.

6 Contracts, Certificates and Notices

Many documents will together make up the building contract (drawings, bills of quantities, specifications, etc.). Of these, the conditions of contract/contract conditions ('conditions') are extremely important, not because they tell either party what the contractor has to build, but because they set out rules agreed between the parties as to how and when the work is to be done, and what payment the contractor will receive and when. The conditions also deal with matters such as terminating the contract, dispute resolution and insurance. The conditions require in many instances that the architect should *certify* or issue a *certificate*, or that either party to the contract should give written *notice*, regarding certain matters relating to the administration of any project. Some certificates and notices will be *condition precedents* to trigger further action on the part of one or both contracting parties. As in other chapters the basis of this chapter will relate to the standard Joint Contracts Tribunal (JCT) 1998 conditions of contract. The general requirements and implications of the more fundamental certificates and notices required are presented.

Contract Commencement

From the contractor's point of view the first indication of what the conditions are will be a reference to either a standard set (such as JCT) or a bespoke set included in the initial tender package. This package will set out in detail the conditions to be used in conjunction with the project and will also detail the amendments to be made, together with contract-specific details such as the amount of the retention fund and the level of liquidated and ascertained damages applicable to the project. These project-specific items will usually be tabulated in a standard format included within the conditions known as the *'appendix'*, which, in the case of JCT 1998, appears towards the back of the conditions. The contractor must note well these provisions and make allowance within the tender price to account for them. It is at this juncture, before spending time and valuable resources on the estimate, that the contractor should decide whether the conditions of contract are acceptable and whether to proceed with the estimate for the project. If there is any doubt with regard to any condition or amendment to the contract then the contractor must immediately query and seek clarity of the amendment before making the tender bid. Qualifications to the original tender at the time of submission will generally lead to disqualification of the contractor's tender.

While there will inevitably be a stringent checking procedure at tender time of the conditions of contract it is equally, if not more, important to check thoroughly through the conditions just before execution. Initially, all contract documents should be checked against the original tender documentation to ensure that no further amendments or alterations have been made to drawings, bills of quantities or conditions. This checking procedure will generally be a function of the contractor's quantity surveyor. The checking procedure should ensure that the appendix has been completed fully and correctly and that all appropriate alterations have been made within the text of the contract. In particular, the contractor's quantity surveyor should check that:

- the contracting parties are as originally stated, and that the client's name is that stated within the original tender documentation. The reasoning for this check is that many developers will set up new companies specifically for the project without any tie to the normal developer's operation. The contractor must confirm at this juncture that funding is in place and that the developer is not a 'man of straw';
- the insurance provisions have been amended correctly and that the client's and/or the contractor's insurances are sufficient to cover the contract requirements;
- the payment terms have not been amended. The contractor must ensure that the payment provisions are as tendered, bearing in mind that subcontractors should have been advised of the original payment terms. One obvious point to note is the payment period and that it has not been amended;
- the contract drawings or schedule of same are all as listed in the original tender documentation. If they are not, then the necessary amendments should be made, or provision must be made for any amendments shown, e.g. to price or programme;
- the rate of liquidated and ascertained damages included within the appendix is in accordance with the original tender enquiry. On occasions this provision may be left blank, but contractors need to be wary of this situation, as it is not the utopia that it seems. The principle is that liquidated and ascertained damages are a genuine pre-estimate of what the employer will lose if the contractor is in default with regard to completion. Although the figure should not and cannot be a penalty, it is there to prompt the contractor to complete the project by the date agreed. If there are no damages stated, that is not '*nil*' damages, but if the appendix is simply left blank, then the contractor will be liable for unliquidated damages, i.e. the actual loss suffered and proved by the employer. This may result in a much greater claim from the employer if the contractor fails the complete on schedule. However, if, for example, the project programme stated within the contract documentation appears to the contractor to be too short, and liquidated and ascertained damages are stated as a set sum, the contractor has the option of making an allowance within the tender to cover any anticipated overrun;
- the period of interim certificates remains as originally tendered, which in the event of the appendix being left blank is one month;
- the retention percentage has been correctly inserted in line with the original tender documentation;
- the date of possession and the date for completion have been correctly inserted in the appendix. On many occasions these dates will differ from the original targets set during the initial tender period and must, by agreement with the employer, be amended accordingly or the price and programme will need amending.

While there are other sections that require careful scrutiny, these are the main areas that must be checked by the contractor's quantity surveyor. In particular, the dates for possession and completion and the amount of liquidated and ascertained damages need checking, especially if the sectional completion supplement section is being incorporated within the conditions of contract.

Letters of Intent

Before the execution of formal contracts, it is not unusual for projects to start using a *letter of intent* as the authority allowing the contractor to commence work on site or, more usually, to allow the contractor to place all necessary orders (or undertake any contractor's design) to allow them to start work on site on the date of possession. This letter of intent may cover the whole of the contract works or it may take the form of a guarantee of payment on a limited expenditure basis. Piling and steelworks often fall within this category, where subcontractors' or suppliers' orders need to be placed so that the subcontractors' internal programming has enough time to incorporate the contractor's programme within their business programme.

There are many views on the acceptability of such documentation. It is up to the contractor to decide whether or not they are comfortable with the document and what protection the letter of intent gives to the contractor should contracts not be signed and the project is aborted.

To overcome such uncertainties the contractor needs to obtain as much security from the letter of intent as he can. As such, the contractor will want the letter to be a separate contract or as near to such a legal status as possible. Conversely, the employer will wish to phrase the letter so that it has as little legal effect as possible. Accordingly, much care needs to be taken to ensure that the letter of intent is not harmful to the contractor in the event that a formal contract is never executed but a substantial amount of work has been done.

The document must come from the employer to be named in the conditions. An instruction from the architect to proceed is not sufficient; as no contract is yet in place, the architect has no authority to issue instructions to the contractor.

The letter of intent should describe in as much detail as possible (usually by reference to the tender documentation) the obligation of the parties, e.g. which conditions of contract are applicable and the extent to which they are to be amended.

It is possible for an entire project to be completed on a letter of intent. If a dispute then arises, for instance, regarding payment, the terms of the letter will be studied carefully. If the letter, in effect, is a contract then it will be construed as any other contract. If, however, it is *not* a contract then the obligations of the parties to each other are limited; for example, the employer will only be obliged to pay a 'quantum meruit' (the value of the work) and that may not equal the contractor's price or even his cost.

A letter of intent in a format beneficial to a contractor is shown below.

[Employer's Name]
[Contractor's Name]
Dear Sirs,

[Project title]

We refer to your Tender document dated [ ] in respect of the works to be carried out at the above site ('the Works') and to our subsequent discussions and correspondence as scheduled.

[list all documentation and dates of discussions]

We are pleased to confirm that it is our intention to enter into a contract with you for the execution of the Works in the sum of [£ *amount of contractor's tender*]. It is our mutual intention that the Works will commence on [ ] and that the contract period will be [ ] weeks.

The Contract Documentation will comprise the following:

[list all documentation, such as type of contract with the amendments as scheduled within the original tender documentation]

[list specification documentation and all drawings that were included within original tender documentation]

[list details of contractor's original tender offer and any subsequent negotiations]

Notwithstanding that no formal contract has yet been entered into between us, please accept this letter as our instruction for you to proceed with the Works described in the Tender Documentation as amended to the extent necessary to enable the commencement date and the contract period referred to above to be achieved.

In the event that a binding contract is entered into between us, your entitlement to payment for any works carried out pursuant to the above instruction will be subject to the terms of such contract. If for any reason we do not conclude a formal contract with you (and we may determine this instruction at any time by 7 days written notice to you) we shall reimburse to you any costs (including cancellation charges where appropriate) properly incurred by you in complying with this instruction up to the date of such termination together with a mark up of 10% to cover head office overheads and profit.

Please acknowledge your acceptance of the above terms and your undertaking to commence and proceed diligently with the carrying out of the Works referred to in this letter by signing and returning the copy enclosed.

Signed ....
for and behalf of the Employer ....
We acknowledge and accept the terms set out above.
Signed ....
for and behalf of the Contractor ....

If the intention of the employer is only to give specific or limited expenditure authority, the letter of intent must particularise the specific elements included within the instruction. This limitation may take the form of an amount of money that should not be exceeded without further instruction, or the letter of intent may relate only to the preordering of certain materials or services. Whichever route is chosen by the employer, the letter of intent must be specific in its authority and the contractor should not exceed the expenditure authorised within it.

Interim Certificates

As set out in Chapter 3, clause 30 of JCT 1998 gives the contractor the benefit of interim certificates so that payments can be made on account during the currency of any project. Without this provision the contractor would have to complete the whole of the project before any requests could be made for payment (subject to the provisions of the Housing Grants Construction and Regeneration Act 1996).

While it is generally the employer's quantity surveyor who prepares the interim valuations in conjunction (but not necessarily in agreement) with the contractor's quantity surveyor, it is the duty of the architect to issue the interim certificate. The architect's only discretion in the matter is that he has the authority to question whether work has been carried out properly and also whether or not materials and goods have been brought to site unreasonably or prematurely. The obligation on the architect is to issue interim certificates at the period stated in the appendix to the conditions up to and including the end of the period during which the certificate of practical completion is issued. Thereafter, the architect's obligation is to issue further interim certificates as and when further amounts are due to the contractor from the employer, provided the architect is not required to issue an interim certificate within one calendar month of the previous certificate.

Failure by the architect to issue such certification in accordance with the contract provisions creates a breach of contract on the part of the employer in respect of which the contractor's remedy is a claim for damages, and may give rise to a right to terminate the contract and/or suspend performance of work.

The interim certificate is required as a condition precedent to payment, with payment being due to the contractor at the time specified in the conditions after the 'date of issue' stated on the interim certificate. The architect should forward directly to the employer a copy of the interim certificate for payment. Some architects will forward both the contractor's and the employer's copy of the interim certificate to the contractor. If this occurs the contractor must ensure that the employer receives the necessary original paperwork, thus ensuring that payment will be made as scheduled. The monitoring of receipt of certificates and their payment is covered in Chapter 3.

If an interim certificate has been issued correctly by the architect but the employer does not honour it, then this amounts to a breach of contract by the employer, for which the primary remedy is a claim for damages. However, two other options are also available.

The first is that the contractor can issue a notice of intention to suspend the performance of his obligations under the contract if the employer's default in payment continues for seven days after the notice has been served by the contractor. This notice to the employer should be copied to the architect stating the grounds on which it is being issued. There is no requirement within this provision for notice to be given by registered post or recorded delivery, but given the magnitude of such action the contractor would be wise to expedite the recording of delivery of such a notice. A simple signature from the recipient acknowledging receipt of the correspondence, showing the date and letter reference, should suffice.

The second option open to the contractor is that of determination of their employment under the contract. JCT 1998, clause 28, is the condition that affords the contractor this right. The notice within the provisions of clause 28 must be given in writing and given by actual delivery, special delivery or recorded delivery. This notice applies to both non-payment of certified amounts and cases where the employer interferes with or obstructs the issue of an interim certificate due under the contract. This provision is an extension to and continuation of the suspension provisions indicated above. The requirement is that, provided the employer's default continues for 14 days from receipt of the original notice of suspension, the contractor may issue a further notice on or within 10 days from the expiry of that 14 day period to the employer determining the employment of the contractor under the contract. Determination takes place on receipt by the employer of such notice.

If the employer rectifies the initial specified default but then repeats the default, the contractor can, within a reasonable time after such repetition, issue to the employer another notice of determination of the employment of the contractor under the contract, again with such determination taking effect from the date of receipt of such further notice. There is no requirement for the contractor to repeat suspension notices in such situations.

To avoid doubt, JCT 1998 sets out in condition 1.8 how the specified dates should be calculated. The contract states that the period of notice shall begin immediately after the notice date and that the notice period shall exclude any public holidays.

The proviso to the above determination procedure is that such notice should not be given unreasonably or vexatiously.

Certificate of Practical Completion

Clause 17 of JCT 1998 contains the provision within subclause 17.1 that when, in the opinion of the architect, 'practical completion of the works' has been achieved, he shall forthwith issue a *certificate* to that effect and that practical completion shall be deemed to have taken place on the day named in such certificate. The issuing of the certificate of practical completion by the architect is an important milestone on any project and triggers many events.

Despite the importance of practical completion to all contracting parties and associated consultants there is no definitive explanation or definition of when practical completion occurs, although it is quite open for such a definition to be incorporated into the contract. Many consider that practical completion occurs at the point of substantive completion, i.e. less than fully complete, but when the building is available to the employer for beneficial use. Unless the contract has been amended, however, the safe view is that the architect will not and should not be expected to issue the certificate of practical completion until the works have been completed entirely in accordance with the contract (i.e. drawings and bills of quantities). One missing door knob is sufficient to allow the architect quite properly to refuse to issue the certificate. Normally, however, the employer will wish to take over the building and the architect will be persuaded to issue the certificate with a 'snagging list' attached.

The decision in the matter of practical completion rests entirely with the architect. Clause 17 does not provide for any consideration to be given to the opinions of the contractor or the employer as to the issue of the certificate of practical completion. There is no requirement within the conditions of contract for the contractor to inform the architect when they consider practical completion has been achieved, although it is usual for the contractor to do this as a trigger for an inspection. In the case of a dispute, it would be sensible for the contractor to record formally that in their opinion the works are practically complete.

Once the certificate of practical completion has been issued, the contractor's quantity surveyor must attend to the following commercial effects.

Clause 30.4.1.3 provides for a release of half the retention relating to the total amount of work that has reached practical completion. Upon the issue of the certificate of practical completion the contractor's quantity surveyor should ensure that a further interim certificate is raised by the architect to include this part release of the retention fund. The timing of the issue of the further interim certificate still falls within the provisions of clause 30.

The contractor's quantity surveyor should ensure also that half of the retention is released to the subcontractors involved on the project.

The contractor's quantity surveyor should consider the position if the *date of issue* of the certificate of practical completion does not correspond with the project date for completion stated in the appendix to the conditions. The reasoning behind this check is dealt with later, in the 'Extension of Time' section. However, a point worth noting here is that the certified date of practical completion marks the end of the contractor's liability in respect of liquidated and ascertained damages if the project has run over schedule and a non-completion certificate has been issued.

The certificate of practical completion also has the effect that it terminates the contractor's obligation to insure the works under clause 22A, 22B.1 or 22C.2, whichever is applicable. To avoid doubt the contractor should notify its insurers that such an event has occurred. In addition, if the contract originally included the requirement for a guarantee or bond to be issued by the contractor then, provided the bond wording is appropriate, i.e. that the bond should be released on the issue of the practical completion certificate, the contractor's quantity surveyor should apply for its return. Many employers will require that they retain the original contract bond document. Cover will generally only cease when the original document is returned to the contractor's surety. It is therefore important to the contractor to ensure the timely release of this document, as costs for its placement will continue to accrue until the original document is returned.

In terms of valuation, the issuing of the certificate of practical completion also marks the beginning of the period for the final valuation of the project. While this matter is dealt with elsewhere in more detail, clause 30.6 stipulates that the contractor should provide to the architect or client's quantity surveyor all necessary documentation for the purposes of adjustment of the contract sum, together with all documentation relating to nominated subcontractors and suppliers not later than six months after the date of practical completion of the works. The architect or the client's quantity surveyor then has a period of three months to complete the final assessment of the final account figure.

Upon the issue of the certificate of practical completion the contractor could begin arbitration on any disputes that could not go to immediate arbitration during the currency of the project, bearing in mind that most precompletion disputes can be dealt with by adjudication.

The issue of the certificate of practical completion also marks the end of the contractor's obligation in respect of carrying out instructions issued by the architect, except in the case of the rectification of defects during the defects liability period, which commences on practical completion being certified.

As a follow-on or extension of clause 17, *practical completion,* clause 18, *partial possession,* contains the provision for early possession by the employer of various sections of the project. The certification takes similar form to that of practical completion, except that the architect issues a certificate of partial possession. All other matters, insurances, retention, etc., follow a similar pattern to that of practical completion of the whole works. The only proviso that clause 18 has over clause 17 is that the employer must seek the contractor's consent to take early possession, and the contractor must not unreasonably withhold that consent, before any certification by the architect. The contractor needs to be aware of the implications of allowing early completion, particularly when it only applies to sections of the project. Allowing the employer access to areas of the project through sections that are not complete can impede the contractor's progress to completion of the project as a whole.

Clause 23.1 requires on the *date of possession* that the contractor begins the works, proceeds in a regular and diligent manner, and completes the works on or before the *date for completion.* The date of partial possession or practical completion of the works is consequently a separate matter to that of the completion date of the project, although there is a link between the dates in respect of non-completion, extensions of time, and the provision of liquidated and ascertained damages.

As a final note, if the architect does not certify that the works are practically complete when the contractor considers they are, then the contractor can refer the dispute either to adjudication or arbitration (if there is an arbitration clause) or to the court.

Certificate of Making Good Defects

Subclause 17.4 of clause 17 requires further certification by the architect, in that, once the architect is satisfied (again, as with practical completion, it is the architect's opinion that is relevant) that any faults arising as a result of materials and workmanship not being in accordance with those required within the contract documents have been made good, he shall issue a certificate to that effect, i.e. the certificate of making good defects.

As indicated previously, the defects liability period starts at the date of practical completion. While JCT 1998 contains a default period of six months if no other time is inserted, many contracts will contain a defects period of 12 months, or six months for general building work and 12 months for mechanical, electrical and lift installations. At the end of the defects liability period the architect must, in accordance with clause 17, issue a schedule of defective works within 14 days of the expiry of the defects liability period. Failure by the architect to produce such a schedule within the contract timetable could result in the employer losing his right to have defects attended to by the contractor at no cost to the employer.

It must be remembered, however, that, while one of the purposes of the defects period is to allow the employer to retain the balance of the retention against defects appearing, the other purpose is to give the contractor the *right* to return to site and rectify those defects at his cost.

It is important to remember that any defect in the works is a breach of contract by the contractor for which, but for the defects liability provisions, the employer would be entitled to damages, these being the cost to him of having the defects rectified. As always, however, in any claim for damages a party must show that he attempted to mitigate his loss. In this case, mitigation would be to allow the contractor to put right the defects at his cost; hence, the defects liability provisions are merely a contractual mitigation measure.

Items such as making good works as a result of frost damage occurring after practical completion and making good of shrinkage cracks at the junctions of rooms or where plaster abuts timber are not defects under the terms of a JCT 1998 contract. If the architect includes such items within the making good of defects schedule the contractor should confirm in writing to the architect that they do not constitute defects within the contract provisions and therefore should be removed from the schedule. The contractor can agree to carry out these additional items either as a gesture of good will or by agreement that they will be paid additional monies to cover the cost of the work.

In addition, many architects will use the medium of the making good of defects schedule to instruct additional works, and again the contractor's quantity surveyor needs to be aware of such items. Remembering that following the issue of the practical completion certificate the contractor has no obligation to carry out further instructions issued by the architect, the contractor must make his own assessment of the situation and advise the architect accordingly if they agree to carry out the additional works and the price for so doing. Whichever route the contractor chooses to take with regard to additional elements of work, their decision should not affect the requirement of rectifying the defective works as scheduled. The contractor's obligation under the contract provisions is to attend to the defects in a reasonable time. Failure to do so may result in the architect issuing a further notice to the contractor requiring that the defects are completed within seven days, failing which the architect may employ others to complete the works. In this case the employer may deduct the costs associated with the remedial works from payments to the contractor and/or pursue the contractor for any outstanding balance.

Again, obtaining the certificate of making good defects is generally a matter outside the brief of the contractor's quantity surveyor, but just as practical completion triggers the first half of the retention to be released, the issue of the making good defects certificate triggers the release of the final retention monies held on the project and the contractor's quantity surveyor should apply for a further interim certificate to facilitate further payment and similarly clear further payments to subcontractors.

Certificate of Non-completion

Clause 24 sets out the employer's contractual remedy for a breach of the contract by the contractor in not completing the contract works on time.

Clause 24.1 provides that, if the contractor fails to complete the works by the date for completion stated in the appendix to the conditions of contract, then the architect shall issue a certificate to that effect. Clause 24.1 goes on to say that in the event of a new completion date being fixed after the issue of such a certificate, such fixing shall cancel that certificate and the architect shall issue such further certification under clause 24.1 as may be required.

The issuing of a certificate under clause 24 triggers the ability of the employer to deduct liquidated and ascertained damages at the rates set out within the appendix to the conditions of contract, for the period between the project date for completion and the date of practical completion. This deduction may be by way of a reduction to an amount certified or as a claim for payment.

If the contractor fails to complete by the date for completion and does not apply for or receive an extension to the contact period the architect has no choice in the matter. The conditions of contract require the notice of non-completion to be issued and the architect must comply. Failure to do so would be a breach of the architect's duties to the employer in not administering the project in accordance with the contract.

If the contractor does not agree with the issue of the non-completion certificate, the contractor should challenge the adequacy of the architect's review of the completion date and not the issuing of the non-completion certificate. As indicated above, the issuing of the non-completion certificate by the architect is mandatory if the date of practical completion exceeds the contract date for completion or revision thereto. The actual withholding of monies from the contractor in respect of liquidated and ascertained damages resulting from the issue of the non-completion certificate remains the prerogative of the employer. The conditions of contract state that the employer *may* withhold or deduct liquidated and ascertained damages. In the case of contracts with local authorities, the damages will invariably be taken with no room for consideration by the employer, as the audit controls within the authorities will always ensure that, once a non-completion certificate is issued, deduction becomes an automatic exercise within their systems. Some commercial clients may be more sympathetic to approaches by the contractor not to take damages, pending final settlement of matters, but fundamentally contractors need to be more proactive in obtaining extensions of time as and when problems occur rather than relying on the goodwill of the employer.

If further extensions of time to the date for completion are granted by the architect, either by reason of the architect's final review, or as a result of further additional matters being raised enabling the contractor to make a further claim, then, if the revised date for completion still does not match the date of practical completion, the architect must issue a new non-completion certificate. Unless this certification is made before the issue of the final certificate, the employer will not be entitled to the liquidated and ascertained damages from the contractor.

Where further extensions to the date for completion are issued, the employer must repay to the contractor monies previously withheld.

It is worth noting that it is a *condition precedent* that the certificate of non-completion must be issued prior to monies becoming due. It is equally a condition precedent that the employer, not the architect, informs the contractor in writing before the date of the final certificate that he requires payment of, or intends to withhold or deduct, liquidated and ascertained damages. Many employers, however, deduct liquidated and ascertained damages from interim certificates before the non-completion certificate is issued. Should this occur the contractor should request the immediate repayment of the monies prematurely withheld.

The contractor's quantity surveyor must ensure that the architect and the employer have acted correctly in such matters and advise accordingly if the correct notices have not been issued. More importantly, the contractor's quantity surveyor should have taken, or ensure that others have taken, the necessary actions that may have been available before the certificate of non-completion was issued. It is fair to say that all is not lost at this juncture on the part of the contractor, but earlier action may help to prevent the problem from arising. As already indicated, early notification of delays is not only a contractual requirement, but also a sound administration policy on the part of the contractor in obviating the issuing of a certificate of non-completion.

Where the project is overrunning it is, therefore, a prime function of the contractor's quantity surveyor to safeguard the contractor from having a non-completion certificate issued by the architect. The surveyor must ensure that, during the currency of the project, when and if problems arise they are properly, fully and formally recorded to the architect. While the matters can be discussed in detail during site meetings, the records or minutes produced are not sufficient to meet the requirements of the conditions of contract. Clause 25.2 calls for written notices to be sent to the architect as and whenever it becomes apparent that the progress of the works has been or is likely to be delayed. It is therefore essential that these notices are forwarded as required. Seeking and obtaining an extension of time to the date for completion by the contractor is the only successful route to fending off the deduction of liquidated and ascertained damages.

Extension of Time

The purpose of the extension of time clause within the contract provisions is to maintain the employer's rights to deduct liquidated and ascertained damages. If there were no extension of time provisions which deal with delays caused by the design team or the employer then, should there be a delay caused by the aforementioned without remedy to the contractor, the liquidated and ascertained damages provision would be invalid. The employer would then have to prove unliquidated damages (i.e. his actual loss) flowing from the failure by the contractor to complete on time.

Clause 25 is the contractor's route to seeking extensions of time to the date for completion. A list of *relevant events* is scheduled under clause 25.4, upon which the contractor can rely to make the necessary extension of time requests. In brief, the relevant events are:

25.4.1 force majeure or act of God
25.4.2 exceptionally adverse weather conditions
25.4.3 loss or damage occasioned by one of the specified perils
25.4.4 civil commotion, strike or lock-out
25.4.5 compliance with architect's instructions

25.4.6 late receipt of information from the architect for which the contractor has made written application
25.4.7 delays on the part of nominated suppliers or subcontractors
25.4.8 execution of work by the employer not forming part of the contract
25.4.9 delay caused by the government or statutory power which restricts the availability of labour or the contractor's ability to secure material
25.4.10 the contractor's inability, for reasons beyond his control, to obtain essential labour or material
24.4.11 delays on the part of local authorities or statutory undertakers
25.4.12 failure of the employer to give in due time ingress to and egress from the site
25.4.13 deferment of possession
25.4.14 significant increases in work carried out for which approximate quantities were included in the contract bills
25.4.15 delays resulting from a change in statutory requirements
25.4.16 the use or threat of terrorism.

Clause 25 places an obligation on the contractor under subclause 25.2.1.1 that if and whenever it becomes apparent that the progress of the works is likely to be delayed, the contractor shall forthwith give *written notice* to the architect of the material circumstances, i.e. the cause of the delay and where appropriate the relevant events upon which the contractor seeks to rely in making the request for the project period to be extended. The requirement placed on the contractor is to notify the architect of all matters that may affect progress, not just matters relating to a relevant event. The requirement exists so that the architect can, upon receipt of such notice, advise the employer of all matters that may affect progress, not just those that are the responsibility of the employer.

The notice should also include, if practical, particulars of the effect of the delay and the consequences that the delay will have on the date for completion. If these details cannot be given at the time of the notice then they should be forwarded by the contractor as soon as they are known.

The contract also contains the requirement, within clause 25.2.3, that the contractor must give further notices as and when further delays occur, or where circumstances change relating to previous notices, again giving particulars of the delay and the effect on the date for completion of the project. The contractor needs to monitor these delays on a regular basis, analysing their effects as the project proceeds. Failure to do so, or failure to respond within a reasonable time to requests for further information regarding delays from the architect, could result in the architect making a judgement on the extension of time requirement without full information. The resultant award may, as a consequence of the contractor's inaction, be unsound and insufficient to cover the total overrun period, leaving the architect with no option other than to issue a non-completion certificate.

A regular *contract review meeting* with the project team is the ideal forum for this area of the project to be discussed. The contractor's quantity surveyor should draw information from these meetings so that the notices required by the contract conditions can be sent. While the notice can be in the form of a simple, fairly short letter, it is advisable for the contractor's quantity surveyor to begin a document known as a *delay schedule* or *cause and effect schedule* (Figure 6.1). Essentially, this is a schedule that identifies:

- the architect's instruction or other events that have caused the programming problems
- a full description of the effect of the problem
- the time when the delay ensued
- the overall delay to the project
- the relevant event applicable to the delay
- the date when the notices were sent.

These delays, their causes and effects should also be plotted on the contractor's programme. The best way to assist the architect, when he is considering whether to grant an extension of time, is to produce the original tender programme with the original critical path amended as a consequence of the delays.

While it may seem premature to commence production of the delay schedule at such an early stage, any claim, whether it is for time or for money, will only succeed if accurate and proper records are kept. It is always advisable therefore to produce records at the time when the problem occurred, when the appropriate records can be properly taken, in particular photographs, site measurements and site records of labour, plant and material affected by the delay.

Producing a schedule at the end of the project, when personnel may have changed or moved on to other projects, can be a very difficult and time-consuming affair and will delay any thoughts of early resolution to the fixing of a new date for completion by the architect. Asking personnel to go back over old ground will also distract them from their current work and may have detrimental effects on the new project.

The architect then has effectively a 12 week period in which to review the contractor's notice and advise the contractor accordingly. The architect must either fix a new completion date by issuing an *extension of time certificate*, which should identify the relevant event under which he is issuing the extension or, alternatively, inform the contractor in writing that he does not consider an extension appropriate and his reason why an extension will not be granted.

The contractor's quantity surveyor must then consider what action to take next. What were the causes of the delays, and can the contractor recover any costs relating to the delays? Clause 26, *loss and expense*, covers the recovery of additional costs.

Loss and Expense

Clause 26 provides the mechanism for the contractor to recover loss and/or expense suffered as a result of disruption to the regular progress of the works. It is not necessary to show a delay to the completion of the project, nor does it necessarily follow that, because the contractor secures an extension of time from the architect, the contractor is able to recover the cost of the prolongation via clause 26. Many of the relevant events scheduled within clause 25 permit the architect to revise the date for completion, but leave the problem of associated costs with the contractor.

Clause 26, similarly to clause 25, contains a schedule or *list of matters* that affords the contractor the opportunity to seek recovery of what clause 26 describes as

106 Commercial management in construction

Murribell Builders Ltd
Delay Schedule

Contract   Date

Architects Instruction/Clerk of Works Directions	Cause and Effect of Delay and Disruption	Period of Delay to Section or Parts of Works	Period of Delay to Completion Date	Contract Clause Relevant to Delay	Contract Clause Relevant to Loss and Expense	Date of Delay Notice and Particulars	Date of Loss and Expense Notice and Particulars

Figure 6.1 Delay schedule.

direct loss and/or expense: essentially, additional costs incurred by the contractor that he would not recover under any other provision in the contract. This *list of matters* comprises:

26.2.1 the contractor not having received in due time necessary instructions for which the contractor has specifically applied, having regard to the date for completion
26.2.2 the opening up for inspection of any covered work, provided the inspection does not reveal that the problem was of the contractor's doing
26.2.3 any discrepancy in or divergence between contract documentation
26.2.4 execution of work by the employer not forming part of the works or the supply of materials by the employer for the works
26.2.5 architect's instructions issued with regard to postponement
26.2.6 failure of the employer to give ingress to or egress from the site
26.2.7 architect's instructions issued under clause 13.2 or 13A.4.1 requiring a variation or under clause 13.3 requiring the expenditure of a provisional sum
26.2.8 the execution of work for which an approximate quantity was included but was not a reasonable forecast of the work required.

As with clause 25, the instigator of the claim must be the contractor. The contractor must make *written application* to the architect stating that he has incurred or is likely to incur direct loss and/or expense in the execution of the contract. The architect has no authority to consider other or periphery matters outside the contractor's notice. It is therefore of vital importance that the contractor ensures that all relevant matters are recorded correctly within the notice.

The notice must include the reasoning behind the application, i.e. the circumstances or causes of the delays or disruptions, highlighting which one or more of the list of matters the contractor is citing as being relevant to the claim.

If the architect is of the opinion that the contractor has suffered direct loss and/or expense, then the architect will ascertain, or instruct the quantity surveyor to ascertain, the amount of such loss. There are three caveats to the above, these being that:

- the contractor's application was made as soon as it had become apparent that the regular progress of the works was affected
- the contractor can supply information in support of the application
- upon the request of the quantity surveyor the contractor can supply to the quantity surveyor details of loss and/or expense as are reasonably necessary for the quantity surveyor to complete his ascertainment.

It is necessary at this juncture, therefore, for the contractor's quantity surveyor to produce the appropriate paperwork together with all necessary supporting documentation.

Final Certificate

The issue of the final certificate is dealt with under clause 30.8 of the conditions of contract. This condition states that the architect shall issue the final certificate not later than two months after whichever of the following occurs last:

- the end of the defects liability period
- the issue date of the certificate of making good defects
- preparation of the final account by the quantity surveyor, which should include any assessment for loss and/or expense.

To ensure that the final certificate is issued as expeditiously as the contract allows, the contractor's quantity surveyor must ensure that all matters within the surveyor's brief are attended to promptly. Looking at the above three parameters, the first item is a matter of fact; the date for the end of making good defects is stated in the appendix to the conditions of contract. The second item, the actual issuing of the certificate of making good defects, is a matter for the architect to deal with, but the contractor's quantity surveyor should prompt the contractor's manager into completing the actual making good work required in a timely manner. Lastly, the final account comes very much within the brief of the contractor's quantity surveyor and is always best dealt with as soon as possible after contract completion, while memories are fresh and all personnel involved are still available.

The conditions of contract require, within the parameters of clause 30.6, that the contractor, i.e. the contractor's quantity surveyor, should provide the architect or the quantity surveyor, not later than six months after practical completion of the works, with all the necessary documents for the purposes of completing the final account in terms of all variations, including the adjustment of prime cost and provisional sums and any adjustment necessary arising out of a claim for loss and/or expense.

If the contractor's quantity surveyor's files, particularly the architect's instructions/variation account file, have been kept up to date and agreement has been reached with the quantity surveyor on a regular basis, then complying with the contract requirements should not be difficult.

Similarly, it is a contract requirement of the quantity surveyor to complete the whole of the final account not later than three months after the receipt of all information. Therefore, the final account should be produced by the quantity surveyor no later than nine months after practical completion of the works, and the final certificate should be issued two months later, provided that the making good of defects certificate has been issued.

There is no requirement in the contract for the final account to be agreed. It is preferable for it to be agreed, but it is entirely correct for the architect simply to issue the final certificate on the basis of the figures provided by the quantity surveyor. If the contractor does not agree with the amount certified he has only a very short time in which to issue a notice of arbitration, failing which the amount certified becomes binding upon him.

Equally, there is no obligation upon the contractor to sign off the final account, and the architect cannot use any such refusal by the contractor as a reason not to issue the final certificate so that any undisputed balance due to the contractor can be paid by the employer.

Recommended Further Reading

Aqua Group (1996). *Contract Administration for the Building Team*. Blackwell Scientific.
Dickason, I. (1982). *JCT 1980 and the Builder*. Chartered Institute of Building.

7 Teamwork and Partnering

Partnering is undoubtedly the biggest buzzword the construction industry has seen in many a year. It is a prime agenda item within most construction-related companies, but what exactly does it mean? Is it just the latest vogue, or is it really a workable ethos within the construction industry, an industry that since the 1980s has become more and more confrontational in its quest for business success? Some will say they have been partnering for years, others will wonder what all the fuss is about, and yet others will have a deep mistrust of working with others and showing their hand. Whatever the viewpoint, partnering is here to stay, certainly for the foreseeable future. It is a workable option: working together as a team and in partnership will achieve much better results than the alternative of conflict and division.

In the construction industry there are two types of partnering, *project* and *strategic*. Project partnering is adopted by many as a trial before the longer term relationship of strategic partnering. This relationship is generally short term, being restricted to one project, and while it can generate savings and improvement for the client and the other team players, undoubtedly the major benefits will come from the use of strategic partnering, where organisations can build the trust and establish the resources required over a period of some years.

There are many definitions of the term 'partnering'; perhaps one of the better ones is: 'two or more organisations working together to obtain mutual benefit and towards a common goal'. Both Sir Michael Latham in 1994 and John Egan in 1998 have proposed the way forward to improving the performance of the construction industry in the UK. Conflict within construction, brought about mainly as a result of the fragmented and adversarial nature of the industry, brought about the need for the government and clients to act and to seek change in the industry. The two leading and most publicised reports of Latham, 'Constructing the Team', and Egan, 'Rethinking Construction', have been used widely to promote initiatives and innovation, with the Housing Grants, Construction and Regeneration Act 1996 setting the scene for major changes throughout the industry.

The need for change within the construction industry is paramount to its survival. Those who fail to recognise and act within this scenario will struggle to succeed. The need for all disciplines to work together as one team, to mutual benefit and in the interests of the client is an absolute necessity. In practice, however, what does this mean?

The success of working within a highly motivated team to a single goal is readily evident in other industries. Nissan's team approach within the motor industry, to incorporate their *just in time* policy with the supply chain, has proved tremendously successful, but the teamwork approach is particularly demonstrable within the sports industry. Whatever the discipline, whether it is rugby, football or rowing, the results of pulling together are very evident. Perhaps a footballer can hide and rest to some degree during a game, but as a very simple analogy imagine Pinsent, Redgrave and co. rowing down the Olympic lake with one of the four pulling at a slower rate; it need only be slightly different and the result would be disastrous. The race would be lost, as would the gold medals. The effect of one of the team not pulling would be evident to all those watching, and therefore the input of the *whole team* at the optimum rate is paramount to the success of the team.

So how did Redgrave and co. achieve their huge success? By *working as a team* in *partnership* towards the common goal. The commitment was obvious to all: extremely hard work, dedication and practice allowed the team to conquer the best in the world. We need also to remember the help and support given to the team, a vital ingredient in any successful team, the rest of the resource, the coaches and training staff, the core team's families, and all the auxiliary help and assistance. Gary Player, one of the all time great golfing professionals, once said, 'the more I practise, the luckier I seem to get', the implication again being that more input fosters greater rewards. This scenario relates to any team or partnering situation.

Within any organisation, therefore, teamwork is absolutely essential to its success. Not only must all departments pull as one, but the whole team must act in unison, drawing upon the total team resource, not just the key players. With partnering this should not be restricted to the different departments within the contractor's organisation, but needs to go a stage further and should embrace all of the people involved in the project, starting with the employer or client and working right through the *whole supply chain*, drawing on relevant expertise as required. The supply chain comprises everybody who can improve the project, from the client right through the chain to the suppliers and their manufacturers. The whole of the team needs to interface with each other if the team is to succeed, which requires huge commitment and dedication from each team player and must include the appropriate support and back-up from all organisations.

Working within a partnership also requires a further main ingredient, and that is *trust*. Without this honest, open approach and commitment to trust within the team, the team would be prevented from achieving the full benefits of teamwork and, as a consequence, may not achieve the final objective of a fully satisfied client. The team will need to engender this trusting philosophy throughout its members, to make it a culture within the group rather than an event that occurs sporadically throughout the life of any particular partnership or individual task.

Working together in partnership in many instances requires a totally different approach or thought process within many organisations. The lack of a trusting culture within the industry, which has developed over many years, needs to be eradicated and converted into the team working scenario. This process requires the utmost commitment from the whole team; it will not occur overnight, nor will it occur as a result of a couple of social meetings. Partnerships require a close relationship, but not a cosy relationship. They are not formed in pubs over a pleasant meal and a glass or two of wine; they evolve out of hard work, a total commitment to each other and a dedication to move the process forward. They need to develop over a period of time, as trust and unity build between parties and, very importantly, the team needs to believe in the principles of partnering.

Building the team requires early participation from all those involved. Traditionally, however, in the construction industry the contractor does not become involved with projects until they are well down the line with design and costing. With a traditional tender the client, the architect and the quantity surveyor will all be well aquainted with the project before they are put out to tender to the selected contractors. By this time it is generally too late for the contractor to offer their knowledge of buildability into the scheme; the design is set and costings have been approved with the client.

In the above scenario both the design and costings will have been completed using a limited number of people and, no matter what their collective experience and expertise has been, they will have missed a valuable resource available to them, that of the contractor's team. The contractor will generally have considerable experience of buildability and costings within their organisation. They will normally have experience of other similar projects for other clients, as will the architect and quantity surveyor, but the contractor can also bring his 'downstream partners', i.e. the subcontractors and suppliers, into the team environment. These additional team members can also bring their respective suppliers and manufacturers into the team, creating a much greater database or resource from which ideas will flow. By using the whole of the supply chain in this way, this enhanced team will bring to the table an ability to discuss all aspects of buildability, new products and innovation, to the betterment of the final product. Using this satellite approach, by sending the problems or thought processes out to the experts, then drawing the results back into the core team, the team will gain the full benefit for the improvement of the project. To obtain maximum benefit, the team must 'look outside the box' and use all of the additional resources available to them.

As indicated at the start of this chapter, if strategic alliances or partnerships are forged, then trust will develop and the benefits will increase. This will be of particular benefit if a standard or repetitive product is being developed.

To focus the way forward five *'drivers of change'* are included in Egan's 'Rethinking Construction', which set out the key factors that need to be addressed:

- focus on the customer
- quality-driven agenda
- committed leadership
- commitment to people
- integrated process and team.

For partnering to be of benefit there is a need for improvement and a need to show betterment. Continuous and demonstrable improvement therefore needs to be achieved. To that end, Egan's construction task force has set certain targets:

- capital cost: reduction of 10%
- time: reduction of 10%
- defects: reduction of 20%
- accidents: reduction of 20%
- predictability: increase of 20%
- productivity: increase of 20%
- turnover and profits: increase of 10%.

To generate this process, early team formation is essential. The earlier the whole team is brought together the greater the resource will be and the greater the result will be. To facilitate this situation an initial workshop or series of workshops should be convened to engender the team philosophy and plot the way forward, to establish the final goals and team objectives and how they are to be achieved. From these initial workshops a *core team* should be established. This core team should comprise key personnel with the authority to act and make decisions on behalf of their respective organisations and should, in general, be restricted to one person from each participating group. There should still be flexibility within this core team. If, for example, legal matters such as condition of contracts, consultant appointments or funding are to be discussed, then it would be entirely appropriate to include within the core team legal representation from those parties involved. In a similar way, if a particular construction method or a particular specification item is to be discussed, then the appropriate expertise should be sought and brought into the core team to gain fully from the available resource.

The core team should be responsible for main decision making, monitoring the team approach, problem solving and dispute resolution. Initially, it should establish specific project objectives, setting targets against which their performance can be measured. The objectives may be in the form of a legal agreement, which would be in addition to the main contract terms, or they may be a set of non-contractual mutual objectives that the team will establish and commit themselves to, for the benefit of the project. These objectives can be as many or as few as the team considers appropriate and may include such items as:

- to deliver the project by the target completion date
- to deliver the project within the agreed target cost and share the financial rewards
- to deliver the project with consideration for life-cycle cost and highest or best value compatible with the quality and level of service required
- to deliver the project with 'no surprises' throughout the life cycle of the project, including the maintenance period
- to ensure minimum disruption to the tenants/occupier or adjacent businesses during the currency of the project
- to maximise the use of local resources and expertise as appropriate
- to deliver the project to the quality targets or key performance indicators set by the partnering team
- to deliver the project safely to the very highest of safety standards.

In addition to these project-specific objectives, the partnering team should consider a set of general partnering objectives such as:

- to provide a reasonable profit for all parties and a relationship that manages risk appropriately, encourages ideas and innovation and rewards excellence
- to achieve a non-adversarial relationship with 'no surprises' through the establishing of effective communications and co-operation between the team partners
- to minimise complaints and enhance reputations
- to use and maintain the best quality resources available to all partners

Project		LOOK-AHEAD SCHEDULE			Date	
Ref	Action required	Champion	Action required by	Action complete		Comments
1	Schedule action points	Name of team member who is to take responsibility for action	Date when task is to be completed	Date when task is completed		Further details as required
2						
3						
etc.						

Figure 7.1 Look-ahead schedule.

- to establish a working culture between the team partners with mutual trust and shared values and objectives
- to strive constantly for continuous mutual improvement in all areas, through the application of effective business management, excellent design and integration with each other.

To apply those matters discussed within the core team, or indeed any other project review, it is necessary to record all matters discussed and schedule these items into sectors either by date or by priority. The *look-ahead schedule* (Figure 7.1) must record the element requiring action, the action needed to achieve completion of the task, a date for proposed completion of the task and the person who is to champion the action point. Each element should be referenced for ease of identification and each element should have the facility to record the date when the task is completed.

In addition to establishing the core team, it is necessary to form a *resource team*, which will be empowered by the core team to implement the day-to-day running of the project. In setting this two-tier approach the core team can concentrate their efforts on the general management of the project, with the overriding objective being the successful completion of the project on or before the target completion dates and on or within the target cost, leaving the resource team to deal with the detailed administration of the project.

To ensure that targets are being achieved and to meet the requirement of verifiable continuous improvement, as already indicated, performance needs to be measured on a regular basis. To facilitate this measurement a series of *key performance indicators (KPIs)* needs to be established.

The core team must decide on this measuring procedure and also decide against what criteria the project is to be monitored. Is the partnership to be measured against standard figures produced by the Department of the Environment, Transport and the Regions (DETR), or does the core team need to establish project- or client-specific KPIs against which to monitor the project?

The DETR identifies 12 factors that can be used to monitor performance. These definitions cover many aspects of the building process.

- Client satisfaction – product   How satisfied was the client with the finished product?
- Client satisfaction – service   How satisfied was the client with the service of the consultants and the main contractor?

- Defects — What was the condition of the product at the time of handover?
- Safety — How many reportable accidents occurred per 100,000 people employed during each year?
- Predictability – cost:
  - design cost: actual cost at 'available for use' less the estimated cost at 'commit to invest'
  - construction cost: actual cost at 'available for use' less the estimated cost at 'commit to construct'.
- Predictability – time:
  - design time: actual duration at 'commit to construct' less the estimated duration at 'commit to invest'
  - construction time: actual duration at 'available for use' less the estimated duration at 'commit to construct'.
- Construction time — The normalised time to construct a project compared with the normalised time to construct a similar project during a previous period.
- Productivity — Assessment of company value added per employee.
- Profitability — Profit before tax and interest as a percentage of sales.

The above criteria are measured in differing ways, some using a scale of 1–10 and others as a percentage improvement over the original comparison figures. Using this methodology comparisons can be made between general information obtainable from the DETR, i.e. comparing against the industry norm. In a similar way, the KPIs can be used to compare project-specific performance within an organisation either with overall company performance or against other projects completed previously. The comparisons can be drawn on a yearly basis or at whatever time the team feels appropriate. Whichever methodology is adopted, the key requirement is to monitor and measure performance to ensure that assessments and improvements can be made.

Unless there is a requirement to measure against DETR figures there is no reason why company- or project-specific KPIs cannot be established, as there is no real need to stick rigidly to government-produced criteria. For example, in a contracting organisation the assessment of design criteria such as the measurement of design time against the original target time may not be required and therefore would not form part of the contractor's KPI assessment, unless the contractor was involved in the design and build process.

Software packages such as 'Contrack Benchmark' are available to facilitate the development of such KPIs, and KPIs can be developed in any organisation by designing simple charts and questionnaires. The essence of the KPIs is to monitor performance against a set of benchmarks so that improvements can be made to the organisation's or team's performance.

To develop the KPIs the organisation or team must decide which areas require monitoring. These criteria may be specific to either a project or the organisation.

The timing of measurement must also be considered and determined. Some measurements will need to be made monthly, whereas others may require only periodic assessment and some can only be assessed at the end of a project or a particular phase within a project.

The organisation or team must also devise the marking criteria for each established KPI and give guide notes to assist the marking team in the assessment. Unlike the DETR indicators, many marking systems allow only a narrow band of assessment, e.g. good, middle or bad, leaving the assessors with little room for manoeuvre and resulting in most assessments being placed in the middle band. In general, there will be two categories of KPI, subjective and objective, but even where the category is subjective, the marking criteria are best geared to give the marking team objectivity within the guidelines for assessing the KPI.

Once the KPIs have been established, whether they are for an organisation or are project specific, the core team should assess and set the targets that are to be the norm, from which assessments can be measured. Careful deliberation is required at this juncture: the target score should be set at a level that can be improved upon, but not at a level that is wholly unacceptable in the first place. The target score should also take account of familiarity with the criteria, e.g. if the KPI is measuring how the team is performing with the use of a new type of contract, then perhaps the agreed target level should initially be set lower than the target level for the measurement of safety elements, where the goal has to be at the highest of levels with a corresponding high target score within the safety KPIs. By way of demonstration, the following KPIs show how assessments can be established.

Many partnering projects are procured using the ECC form of contract, a contract new to many contractors who traditionally have been more used to the JCT family of contracts. It may be the intention of the team to measure how they perform against some of the requirements of the contract; for example, ECC has a requirement of *'early warning'* of *compensation events* (variations). The KPI in Table 7.1 could therefore be established to assess this area of the team's performance.

For this particular KPI, if the contract is new to the team, then the target level may be set at a relatively low but realistic level of, perhaps, level 3. If the team is experienced with the ECC contract and fully understands the requirement for early warning and methodology required for recording the early warning of compensation events, perhaps the target should be set higher, at level 6. In this case,

Table 7.1 Early warning: compensation events (CEs)

Score	Criterion
0.00	Not scored
1.00	0% CEs notified with early warning
2.00	Up to 10% CEs notified with early warning
3.00	Up to 25% CEs notified with early warning
4.00	Up to 40% CEs notified with early warning
5.00	Up to 60% CEs notified with early warning
6.00	Up to 75% CEs notified with early warning
7.00	Up to 90% CEs notified with early warning
8.00	All CEs notified with early warning

the target is set at a high but sensible level that can still be improved upon with a little more effort from those involved.

With other KPIs, such as safety and defects at handover, the target must be set high, as poor performance in these areas is simply not acceptable.

Within the DETR standard KPI assessments, only reportable accidents are monitored. These are defined by the Health and Safety Commission as being fatalities, major injuries and injuries to employees causing three days or more absence from work, this definition applying not only to employees but also to the self-employed and members of the public. However, both reportable and non-reportable accidents can be assessed. Safety must be the highest consideration for any organisation or project team, with constant vigilance and improvement being necessary. Table 7.2 indicates how these can be measured and how the target level is set very high but still allows for improvement.

This table can be used for both criteria with the target being set high at 7, but still allowing for continuous improvement by the team. The table is also set out in such a way as to weight the scoring criteria heavily towards success in achieving zero accidents.

Whichever method is required or adopted by the team, the KPIs should be established in sufficient time to facilitate early measurement of performance. In addition to the agreement of the target score for each KPI, the team must establish who will supply the information to be used in the scoring process. For example, on project-specific KPIs, the contractor will be best suited to supply safety records of reportable and non-reportable accidents, the project manager will be in a position to advise on early warning of compensation events, and the employer or client will need to input when it comes to the assessment of satisfaction with product and the project team's performance. The process as a whole must, however, remain a team operation, with all members inputting into the assessment process and, more importantly, with all team players being advised of the results as they are assessed, so that action plans can be put into place as appropriate to aid the process of seeking continuous improvement.

The above process of working together within the *partnering* scenario should not be restricted to organisations working together or to projects procured on a partnering basis. The key criterion is that of a *team* approach, working together, monitoring performance and seeking improvement wherever possible.

Table 7.2 Zero accidents: reportable and non-reportable

Score	Criterion
0.00	More than 30 accidents in the period
1.00	24–30 accidents in the period
2.00	18–23 accidents in the period
3.00	12–17 accidents in the period
4.00	7–11 accidents in the period
5.00	3–6 accidents in the period
6.00	1–2 accidents in the period
7.00	No accidents in the period
8.00	No accidents during the year or contract period

Taking the Team to Site: The On-site Process

The above section, especially the measurement of project-specific performance, will normally be carried out at some distance to the on-site process, although the bulk of the measuring criteria will be of site-related matters. Similarly, the lookahead schedules will be developed as the vehicle for moving the project forward, with the bulk of the requirements being actioned before starting on site. This scenario should not, however, distract from developing the team approach right through the building process, between colleagues, between organisations and, very importantly, between site personnel, site operatives and office staff. Team work and team building are the key factors; they need no buzzwords, just hard work, hard work and more hard work. The need to keep the site operation as an integral part of the team is paramount to success. The frequency of the core team's meetings may be changed to monthly as projects develop onto the on-site situation. The resource team will then take over the day-to-day operation process, but this is not the time to revert back to more normal individual roles. Far from it: the new team now needs to work to build the trust that will move the project towards a successful and rewarding conclusion. This applies equally to any organisation as it would to individual projects. There will be a management structure in place, and the core team may comprise the directors or partners, but there will also be management teams and different departments, all of whom need to develop a team approach.

To keep the motivation going, the resource team needs to establish the way forward, via a series of team meetings. This scenario applies equally to projects running on a traditional procurement basis. Traditionally, this has been carried out using a series of site meetings chaired by the architect and involving the architect, quantity surveyor and contractor, these meetings generally being held on a monthly basis. With many of these meetings, progress is the key factor, although information required will also form part of the discussion; technical matters and costings, however, are generally taboo, as is direct input from subcontractors and suppliers. The assumption is that such matters should have been discussed elsewhere, outside the formal site meeting. However, it may be beneficial to involve the whole team in the process. Although there is a need to prevent meetings for meetings' sake and to prevent meetings from becoming an unwieldy process, by not using the whole resource the team will lose a major benefit in the operation of the project.

The use of *contract review meetings* by the resource team will help this team approach and draw together all the necessary resources to the betterment of the project. The timing of these review meetings should be decided by the resource team, and will depend on the complexity of the project, but in general two per month should suffice. The resource team should also decide who is required to attend and who are the key players, and manage the timing of the meetings to prevent time being wasted for those who may only need to input into part of the meeting. The key factor, however, should be openness and honesty with each other. Where problems arise, the team members must work together to resolve the situation. They must adopt a *no blame culture*, accepting that they are all responsible for the project and for ensuring a successful conclusion.

These contract review meetings should look at all aspects of the partnership; items such as cash flow, funding and tenant issues should not be precluded from the discussion. The team must remember that each player will have their own agenda with elements which are very important to them and that require review as much as the architect's design process or the contractor-based areas of progress, information requirements and general construction-related elements. There will be a need for the agenda to stick to matters of prime importance to avoid long sessions; therefore, detailed design or technical construction issues should be dealt with outside the main contract review meetings.

The contract review meeting should not simply be a process of recording events that have occurred previously, but should be used as a platform to move the project forward in a positive manner. There is a need to record progress, but the actions required to maintain or improve progress should be the key factors. Similarly, with costings, the current situation will need to be recorded, but the meeting should concentrate on the actions required in value engineering, who is required to do what, who is taking responsibility for the required actions, and when does the task need to be complete. These are the important factors, the key ingredients that will generate the enthusiasm to keep the process going. At both core team meetings and resource team contract review meetings, the team must put in place action points that are considered necessary to bring improvements to the project or process being debated.

Appendix 1

A Worked Example

The following documentation represents an actual interim valuation on a city centre refurbishment project. The example begins with the contractor's quantity surveyor's valuation and works through to the cost valuation comparison for discussion with the management team. The example comprises the following documentation:

- the interim application submitted to the client's quantity surveyor, which includes:
  - preliminaries (Figure A.1)
  - measured work sections (Figure A.1)
  - schedule of architect's instructions (Figure A.2)
  - schedule of sundry variations (Figure A.3)
  - schedule of materials on site (Figure A.4)

- the response of the client's quantity surveyor, which includes:
  - amended contractor's measured account (Figure A.5)
  - amended schedule of architect's instructions (Figure A.6)
  - amended schedule of sundry variations (Figure A.7)

- the architect's interim valuation certificate (Figure A.8)
- the contractor's internal notification of payment due (Figure A.9)
- the contractor's cost valuation comparison, including:
  - two sample subcontract liability schedules (Figure A.10 and A.11)
  - subcontract liability summary schedule (Figure A.12)
  - contractor's internal preliminary schedule (Figure A.13)
  - contractor's valuation adjustment schedule (Figure A.14)
  - schedule of other costs (Figure A.15).

Note that, for the sake of brevity, back-up information relating to the precise pricing of architect's instructions and sundry variations has been excluded from this example.

To ensure privacy all names have been changed throughout the example.

As described in the main text, all information appertaining to the interim application, including analytical pricing of variations, should be enclosed as appropriate with each valuation submission.

Looking first at the contractor's interim application (Figure A.1) the document is well presented, easy to follow and has been produced using a simple scrolling methodology through the contract bills of quantities. Omissions from the submitted application, however, are the name or reference of the contractor's quantity surveyor who carried out the valuation and the date when on-site notes were taken. This information may be recorded elsewhere, but if it is highlighted on the application itself the record is available for future reference by others.

It can be seen from Figure A.1 that the preliminaries have been claimed using the same scrolling technique as the main measured works sections, the quantity surveyor using the originally priced elements in the bills of quantities. As described in Chapter 3, the technique of scrolling the bills of quantities is more than acceptable for the purpose of interim valuations, provided both quantity surveyors are confident that their assessments of percentages of work complete will produce an accurate assessment overall.

In the example shown, the contractor's quantity surveyor has used both percentages and actual measured quantities in the build-up of the application. Where actual measurements are used in the compilation of the application, they should be appended to the application to facilitate checking and agreement by the client's quantity surveyor. This would equally apply whether the measurements were provisional or firm measurement records for later use in the final accounting process.

Included with the contractor's application are assessments for both the valuation of architect's instructions issued to date and sundry variations carried out by the contractor before formal confirmation by the architect (Figures A.2 and A.3). As mentioned, back-up information relating to these areas of work is not included in this example but should be enclosed with any submission made. Working strictly to the letter of the contract provisions, variation work carried out will only be included in interim assessments by the client's quantity surveyor if an appropriate architect's instruction has been issued. In most instances, however, assessments will be included in the interim valuations to cover such works. In this example, the sundry variations are scheduled and valued, and where architect's instructions have been subsequently issued the sundry variation value is transferred to the architect's instruction schedule with an appropriate reference being noted on the schedule. The contractor should not, however, rely on this relaxation of the contract provisions; the contractor's team should endeavour at all times to ensure that the appropriate architect's instruction is issued to cover any varied works.

The next document in the example is the assessment by the client's quantity surveyor of the interim valuation (Figures A.5–A.7). In this example the client's representative chooses not to work through the valuation with the contractor's representative as it is produced, preferring to amend the contractor's proposals as considered necessary. The situation is not ideal, but will be the format encountered on many occasions by the contractor's quantity surveyor. The contractor's preference should be to prepare the valuation and subsequently agree the contents with the client's representative before submission. If the respective quantity surveyors do not meet to agree the interim application then the contractor's quantity surveyor must keep close contact with the client's representative, answering queries as they occur, but more importantly avoiding the situation where the client's quantity surveyor issues an amended interim valuation without discussion with the contractor. In this example, the contractor's quantity surveyor forwards the application to the client's quantity surveyor electronically in Lotus 123 or Excel format, this facilitating quick transfer of information and ease of amendment should the client's quantity surveyor deem it necessary. As can be seen from the example, the client's quantity surveyor has made several amendments to the contractor's application and has highlighted these changes for ease of checking by the contractor. The system works well and allows ease of discussion over the amended areas of the application. This methodology would

Murribell Builders Ltd
City Centre Hotel
Interim Valuation No 5 - 28th July 2000

Contractor's submission

Preliminaries

1/16 - Bond	%	100		444.00
1/63 - Management and Staff	%	95	39,156.00	37,198.20
1/64 - Site Accommodation	%	95	3,429.38	3,257.91
1/65 - Power - additional services	%	95	15,100.00	14,345.00
- Telephone installation	%	100		360.00
- Cleaning	%	20	800.00	160.00
1/66 - Hoist - erect/dismantle	%	50	1,860.00	930.00
- Hire	%	100		8,010.00
- Transport, concrete plant	%	100		910.00
1/67 - Scaffolding	%	95	62,429.00	59,307.55

4th & 5th Floor Alterations

3/1/C1 - 3/5/K4	%	100		24,458.25
3/5/K5-K7	%	75	10,733.74	8,050.31
K/8-K10	%	60	920.79	552.47
3/6/K11	%	60	43.95	26.37
K12	%	100		292.50
K13	%	85	6,316.18	5,368.75
K/15, K16	%	100		2,276.47
3/7/K17 - K23	%	100		7,852.36
3/8/K24 - K33	%	100		7,082.78
3/9/K34, K35	%	100		53.08
L1	No	5	458.15	2,290.75
L2	No	2	262.35	524.70
L3	No	2	139.03	278.06
L4	No	2	82.78	165.56
3/10/L7 - L10	%	87.5	426.18	372.91
L14-L17	%	67	360.46	174.51
3/11/L18	%	67	117.15	78.49
M1	%	100		29.30
3/12/M3	%	27	16,447.79	4,440.90
M6	%	50	6,106.24	3,053.12
3/13/M12	%	50	307.35	153.68
3/14/P1	%	85	1,002.54	852.16
P2,P3	%	50	458.64	229.32
P4	%	90	1,164.24	1,047.82
			c/fwd	194,627.28

Figure A.1

			b/fwd		194,627.28
4th & 5th Floor Alterations (continued)					
3/14/P5 - P7	%	100			8,778.00
3/15/P8-P11	%	50		11,874.92	5,937.46
3/16/P19-P27	%	100			1,015.88
3/17/P28-P32	%	100			3,856.74
Lift Shaft					
4/1/C1 - 4/7/L9	%	100			57,937.34
4/8/M1 - M5	%	100			4,366.55
4/9/P1, P2, P7	%	100			277.02
4/10/P8, P9	%	100			140.30
South Wing Staircase					
5/1/C1-5/4/G3	%	100			5,670.78
5/5/L1	%	1		359.32	359.32
L2 - L4	%	33		167.65	55.32
5/7/P5 - P7	%	100			334.90
People's Gallery					
6/1/C1 - F4	%	100			1,799.76
6/2/L1	%	100			273.63
6/3/L2 - M2	%	100			553.29
6/4/M4 - M7	%	100			775.18
6/8/P5 - P7	%	100			320.30
Roof					
7/1/C1 - C4, C9 - C11	%	100			3,989.89
7/2/C12-C22	%	100			9,566.22
7/3/G1 - G4	%	100			1,157.09
7/4/G5 - G16	%	100			1,814.42
7/5/G17, G21 - G25	%	100			1,011.65
7/6/H1 - H9	%	100			31,387.71
7/7/H10 - H19	%	100			5,959.49
7/8/H20 - H27	%	100			1,839.30
7/9/H28 - H34	%	100			2,170.35
7/10/H35 - H42	%	100			6,932.34
7/11/H43 - H47	%	100			10,174.42
			c/fwd		363,081.93

Figure A.1 (*Continued*)

			b/fwd		363,081.93
Roof (Continued)	%				
7/12/K7. L1	%		100		880.99
7/13/M1 - M6	%		100		233.51
7/14/P1 - P3	%		77	15,275.40	11,762.06
R1 - R5	%		100		322.07
7/15/R6 - R11	%		100		410.89

External Walkways

8/1/C1 - C4	%	100		6,357.61
8/2/C5 - C11	%	100		3,676.17
8/3/C12 - E4	%	100		9,552.14
8/4/E5 - F1	%	100		839.20
8/5/G1 - G5	%	100		12,033.90
8/6/G6 - J4	%	100		9,768.13
8/7/J5	%	100		325.23
M1 - M3	%	50	933.43	466.72
P1, P2	%	100		1,940.50

Mechanical Installation

9/1	%	82.25	94,463.00	77,695.82

Electrical Installation

10/1	%	60.81	105,284.00	64,023.20

Lift Installation

11/1 - see attached invoice. See AI No 2.01	%	100		86,092.50

Architect's Instructions

As attached summary		sum	172,532.05

Sundry Variations

As attached summary		sum	2,468.74

Material on Site

As attached summary		sum	65,800.46
			890,263.82
Less	Retention @ 5%		-44,513.19
			845,750.63
Less	Previously certified No 4		-573,542.00
	Amount Due		**£272,208.63**

Figure A.1 (*Continued*)

Murribell Builders Ltd
City Centre Museum

Architect's Instructions

Interim Valuation No.5 - 28th July 2001

A.I. No.	Item	Description	Omit		Add	
1	1.01	2 copies of tender drawings form contract set	0.00		0.00	
	1.02	2 copies of drawings amended since tender	0.00	03A	TBA	
				04B	TBA	
2	1.01	Lift car 2350mm wide and 2400mm deep	0.00		0.00	
	1.02	Lift doors to have minimum opening of 1300mm	0.00		0.00	
	1.03	Changes on new walls doors and service inst.	0.00		0.00	
	2.01	Contract sum increased by £93,621.09				
		- Ironmongery to extra doors £9,000.00	0.00			
		- Changes to mechanical services £5,000.00	0.00			
		- Changes to electrical services £7,000.00	0.00			
		- Lift installation £72,621.09	0.00		42,690.84	Part only
	3.01	Contract period extended to 23 weeks	0.00		0.00	
3	1.00	2 copies of drawings				
		G3882/1L(--)11b,12b,13d,14c,15b,16b,17b	0.00		0.00	
		G3882/1A(66)01a	0.00		0.00	
		G532/1A(31)01a	0.00		0.00	
		G3882/4/01	0.00		0.00	
		G3882/4/02a	0.00		0.00	
	2.00	Contract start date 6th March 2000	0.00		0.00	
4	1.00	2 copies of drawing G3882/1L(--)03	0.00		0.00	
	2.00	Alterations to screen positions	0.00		0.00	
5	1.00	Provisional Sums				
	1.01	Item Z4				
		Omit sum for window shutters £(14000.00)	0.00		0.00	
		Add works as Bowey quotation £7494.81	0.00		0.00	
	1.02	Item Z7				
		Omit sum for external door £(3,500.00)	0.00		0.00	
		Add works as Bowey quotation £2,321.97	0.00		0.00	
	1.03	Item Z8				
		Omit sum for stair extension £(5,000.00)	0.00		0.00	
		Add works as Bowey quotation £5,866.09	0.00		5,866.09	
		To summary	0.00		48,556.93	

Figure A.2

Murribell Builders Ltd
City Centre Museum

Architect's Instructions

Interim Valuation No.5 - 28th July 2001

A.I. No.	Item	Description	Omit	Add
6	1.00	2 copies of drawings		
		G546/001a	0.00	0.00
		G532/002b,003c,004a,005	0.00	0.00
		G3882/1L(--)12c	0.00	0.00
		G532/1A(76.7)01	0.00	0.00
	2.00	New Lift Shaft		
		Bond blockwork with resin ties every course	0.00	0.00
	3.00	Electrical Queries (CWS Queries No.1)		
		Ref. DY1		
	3.01	Areas out with storage omitted, fire detection		
	3.02	Areas omitted CD to issue revised drawings		
	3.03	FA panel to be resited in Peoples Gallery		
	3.04	No, MICC to be used as specified		
	3.05	Areas omitted CD to issue revised drawings		
		Ref. DY2		
	3.06	Main contractor		
	3.07	Areas to be surface conduit as specification		
		Ref. DY3		
	3.08	Areas omitted CD to issue revised drawings		
	3.09	Existing fire alarm maintained until new in place		
	3.10	Main contractor		
	3.11	Main contractor		
	3.12	Ref. F,F/E lum's as Ref. F1. Ref. A1 as Ref. A.		
	3.13	Redirect electrical services using specialist		
	4.00	Omit linoleum as M50/150A	0.00	0.00
		Add vinyl as M50/1501 for e.o. £11,474.87	0.00 27%	3,098.21
		To summary	0.00	3,098.21

Figure A.2 (Continued)

Murribell Builders Ltd
City Centre Museum

Architect's Instructions

Interim Valuation No.5 - 28th July 2001

A.I. No.	Item	Description	Omit		Add
7	1.00	Provisional Sums			
	1.01	Item Z2			
		Omit sum for bird control netting £(23,000.00)			
		Add works as Bowey quotation £26,284.50	0.00	60%	15770.70
	2.01	2 copies of drawing G352/sk-01a	0.00		
	3.00	Electrical Queries from CWS			
		Ref. DY4			
	3.01	Bowey Construction			
	3.02	Fire alarm panel fed via 20A SP&N			
	3.03	Dark Room area consumer unit as note 2			
	3.04	Areas cannot be left without lighting, use temp.			
		Ref. DY5			
	3.05	Main contractors information			
		Ref. DY6			
	3.06	Awaiting client information			
		Ref. DY7			
	3.07	Awaiting client information			
8	1.00	Electrical Queries from CWS			
		Ref. DY4			
	1.01	E1 luminaire around lift shaft shall be slave unit			
	1.02	E2 units to stairs and lobbies shall be slave unit			
	2.00	Omit disabled alarm panel. Liaise with Shorrock	0.00		1,100.00
	3.00	2 copies of drawings			
		G532/F/01c,02c,03c,04c,05c			
9	1.00	Provisional Sums			
	1.01	Item Z6			
		Omit sum for ironmongery £(17,500.00)	0.00		0.00
		Add works as Bowey quotation £29,684.28	0.00		0.00
		To summary	0.00		16,870.70

Figure A.2 (Continued)

Murribell Builders Ltd
City Centre Museum

Architect's Instructions

Interim Valuation No.5 - 28th July 2001

A.I. No.	Item	Description	Omit	Add	
9 (cont'd)	1.02	Item Z9 Omit sum for removal of rubbish £(1,000.00) Add works as Bowey quotation £2,708.00		2,708.00	
	2.01	Strip out emergency lighting to existing stairwell			
	3.01	Approval of samples (roof tiles, bricks)	0.00	0.00	
10	1.00 1.01	Confirm COW's Direction No.1 Basement lift area - disconnect all cables, etc	0.00	0.00	see AI No.16/2.03
	1.02	Take out and re-route dry riser			
	1.03	Ground floor - remove rad. in front of window			
	1.04	Remove all rubbish from 4th and 5th floors	0.00	0.00	see AI No.9/1.02
	1.05	Fixings for scaffold go into centre of bricks	0.00	0.00	
	1.06	Confirm COW's Direction No.2 Issue of drawings G546t/001a G532/002b,003c,004a,005 G3882/1L(--)12c G532/1A(76.7)01	0.00 0.00 0.00 0.00	0.00 0.00 0.00 0.00	
	1.07	Confirm COW's Direction No.3 Add 45mm isolator for consumer unit in dark rm. Omit 32mm isolator in same			
	1.08	Add suspend fluorescent fittings on gnd. flr.			
	1.09	Add use of new service cupds for cable trays			
	1.10	Add mount section board in service cupbd			
	1.11	Confirm COW's Direction No.4 Build up front of roof gutter 75mm	0.00	4,415.40	
	1.12	Lead gutter lining in three pieces of lead	0.00 Incl.	0.00	
	1.13	Box out gutter to form expansion joint every 2m	0.00 Incl.	0.00	
	1.14	Repair stone gutter as instructed on site	0.00	7,249.13	
		To summary	0.00	14,372.53	

Figure A.2 (*Continued*)

Murribell Builders Ltd
City Centre Museum

Architect's Instructions

Interim Valuation No.5 - 28th July 2001

A.I. No.	Item	Description	Omit		Add	
10 (cont'd)	1.15	Replace 2 sections of stone gutter	0.00		1744.60	as quote
	1.16	Repair stone window cill to north elevation	0.00		4,772.90	as quote
	1.17	Confirm COW's Direction No.5 Omit C22 , Page 7 / 2 Add take off half to inner wall for ventilation	-5,379.72 0.00		0.00 952.50	127m @ 7.50
	1.18	Omit C21 , Page 7 / 2 Add sarking board at eaves for width of 600mm	-556.27 0.00		0.00 333.19	143 @ 2.33
	1.19	Confirm COW's Direction No.6 Repointing all brickwork above eaves level Include all scaffolding	0.00 0.00	267m2	4,170.54 0.00	see 2.03 & 2.04
	1.20	Confirm COW's Direction No.7 Carry out work to gutter as Bowey quote	0.00		0.00	see AI No.10/1.14
	2.00 2.01	Confirmation of faxes sent to site Sketch for position of metal trays for services				
	2.02	Replace 2 No. stone gutters	0.00		0.00	see AI No.10/1.15
	2.03	Erect scaffolding to water tower for repointing	0.00		1,930.00	
	2.04	Erect scaffolding to chimney for repointing	0.00		533.50	
	2.05	Plywood to 4th floor - lay additional polythene	0.00	940m2	921.20	
	2.06	Enlarge DG03 as sketch	0.00		0.00	
	3.00	Issue of drawings G532/1A(31) 01b G532/1A(31) 02a G532/1A(76.7) 01b G532/1L(--) 13e	0.00 0.00 0.00 0.00		0.00 0.00 0.00 0.00	
11	1.00 1.01	Additional works Stone repairs to 4th floor window cills as quote	0.00		0.00	see AI No.10/1.16
	1.02	Remove dormer windows and replace as quote	0.00		6,549.84	
	1.03	Repair area of roof south of valley on west elev.	0.00		356.40	
		To summary	-5,935.99		22,264.67	

Figure A.2 *(Continued)*

Appendix 1: A worked example 129

Murribell Builders Ltd
City Centre Museum

Architect's Instructions

Interim Valuation No.5 - 28th July 2001

A.I. No.	Item	Description	Omit		Add	
12	1.01	Issue of drawings				
		G532/1S(31) 01a	0.00		0.00	
		G532/1S(31) 02b	0.00	30%	217.73	
		G3882/F/01d , 02d , 03d , 04d	0.00	25%	48.44	
			0.00			
			0.00			
			0.00			
13	1.01	G532/1L(--) 11c , 12d , 13f	0.00		1,947.79	Conc floor to lift shaft
	2.00	Confirm COW's Direction No.8				
	2.01	Remove 3sq. m of screed in Dark Room	0.00			
	2.02	Roff lights - cast wire panels damaged	-211.28	74No	1,954.34	7/6/H3
	2.01	Confirm COW's Direction No.9 Repoint all brickwork above eaves level	0.00		0.00	see AI No.10/1.19
	2.02	Repair stone gutters and stone window cills	0.00		0.00	see AI No.10/1.16
	2.03	Extend gutter lead outlets to c. i. down pipes	-243.80	15No.	365.70	7/10/H39
	2.04	Peoples Gallery - cut hole along side of cable				
	2.05	Confirm COW's Direction No.10 Omit continuous roof ventilator to patent glazing	0.00		0.00	
	2.06	Add 50mm holes through sarking board	0.00		0.00	
	2.07	Confirm COW's Direction No.11 Paint screen and door in Turbinia Gallery				
	2.08	Drawing G532/1A/31/01c	0.00		0.00	
	2.09	Copy of fax regarding floor finishes	0.00		0.00	
	2.10	6mm ply. to top and side of dormer windows	-498.40	104m2	459.68	7/5/G19,G20
	2.11	Confirm COW's Direction No.12 Replace timber surround to lift access as quote	0.00		1,605.95	
	2.12	Levelling screed at 4th floor to take out hollows	0.00		972.96	
	2.13	Ply. under vinyl sheet to have fixings punched	0.00		0.00	
		To summary	-953.48		7,572.59	

Figure A.2 (Continued)

Murribell Builders Ltd
City Centre Museum

Architect's Instructions

Interim Valuation No.5 - 28th July 2001

A.I. No.	Item	Description	Omit		Add	
13 (cont'd)	3.00	Confirmation of fax				
	3.01	Lift pit - Omit visqueen tanking and korkpak	-83.30		0.00	
		Add servicised preprufe system	0.00		1,705.05	
		Electrical queries				
	3.02	Order 7.5kVA emergency lighting cubicle				
	3.03	Batteries as original spec., i.e. not plante cells	0.00		5,566.14	50% increase capacity
	3.04	Do not provide earth leakage monitor to board				
	3.05	Upgrade pvc conduits to high impact	0.00			
	3.06	Omit contactor controlled lighting to 4th floor	0.00		0.00	see AI No.16/2.03
14	1.01	Finishes				
	2.01	Omit Rockwool RW3, Add Rocksil RS60	0.00		0.00	
	3.00	Confirm COW's Directions				
	3.01	No.13 - Add 6mm cable at 4th floor	0.00		217.83	
	3.02	No.14 - Powder coating for roof vents	0.00		1,738.00	
	4.01	Omit electromagnetic locks to DG07 & D303	0.00		0.00	incl. AI No.12/1.01
15	1.01	Omit string skirting to stairs, Add sealant		6/5/M12.1		
		Omit coved skirting, Add set in skirting			66m	425.70
		Add butt joint to be pointed with sealant	0.00	23m	37.95	
	1.02	Quote for Gradus Pathfinder nosings	0.00		0.00	
	2.02	Drawing issue - G532/FLOOR-01 rev.A	0.00		0.00	
	3.00	Electrical queries (ref. DY11 & DY12)				
	3.01	CWS will provide containment supplies	0.00			
	3.02	Darkroom - flush to walls, surface to ceiling				
		Dis. wc - flush to walls, surface to ceiling				
		Gflr cor. - flush to walls, surface to ceiling				
		Lift lobbies/corridors - new walls concealed				
		- existing walls surface				
	3.03	DY12 identical to DY11				
		To summary	-83.30		9,690.67	

Figure A.2 *(Continued)*

Appendix 1: A worked example 131

Murribell Builders Ltd
City Centre Museum

Architect's Instructions

Interim Valuation No.5 - 28th July 2001

A.I. No.	Item	Description	Omit		Add	
16	1.01	Omit Flotex carpet mats to lift doors	0.00		0.00	
	1.02	Quote for Gradus ELAL nosing	0.00		0.00	
	2.00	Electrical Queries				
	2.01	Switchboard drawings - no comment				
	2.02	CWS latest lift drawings				
	2.03	CWS variations accepted	-9,929.36		9,309.90	items 2 - 14 ; as quote
	3.01	5th floor ceiling - galv. angle brackets	0.00		0.00	
	4.01	Shower bench - do not order yet	0.00		0.00	
	5.01	External walkways - steelwork	0.00		0.00	
	6.01	COW's D No.15 - studs in column base	0.00		0.00	see AI No.17/2.01
17	1.01	Omit prov. sum for peoples gallery	0.00		0.00	
		Add Bowey Construction quote	0.00		2,237.83	part only
	2.01	Cut back 2No. column bases to lift pit	0.00		705.65	
	2.02	Add collars to cast iron pipe in lift pit	0.00		0.00	see AI No.20/4.06
	3.00	Omit prov. sum for film to dormer windows	0.00		0.00	
18	1.01	Intercom & text writer as quote	0.00	50%	14,293.68	
	2.01	Lift car finishes				
	2.02	No fire rating of lift car				
	3.00	Electrical Queries				
	3.01	DY12 answered previously				
	3.02	DY7 - awaiting client response				
	3.03	Allgoods to supply information to CWS				
	3.04	CWS contacted regarding drawings				
	4.01	External walkways paint approved	0.00		0.00	
	4.02	No comment on welding details	0.00		0.00	
19	1.01	Liaise with Northumbrian Water re: temp. water shut down	0.00		0.00	
		To summary	-9,929.36		26,547.06	

Figure A.2 (*Continued*)

Murribell Builders Ltd
City Centre Museum

Architect's Instructions

Interim Valuation No.5 - 28th July 2001

A.I. No.	Item	Description	Omit	Add	
20	1.01	Metal stair clarification/amendment	0.00	715.00	
	2.00	Handrails to west & south wing stairs	0.00	0.00	
	2.01	Quote for upgrading	0.00	0.00	
	- 2.07	existing and new handrail	0.00	0.00	
	3.01	Handrail to metal stair amended	0.00	0.00	
	4.00	Confirm COW's Directions			
		- No.16			
	4.01	Vinyl sheet to stair is Armstrong Medintech	0.00	0.00	see AI No.6/4.00
	4.02	Vinyl to other areas is Armstrong Tapestry	0.00	0.00	NOT used/claimed
	4.03	Plywood fixings punch nailed	0.00	0.00	as BOQ's
		- No.17			
	4.04	Lift shaft base - break out column bases	0.00	0.00	see AI No.17/2.01
	4.05	Paint 3No. windows to water tower	0.00	0.00	
		- No.18			
	4.06	Drainage in lift shaft - replace c.i. pipe	0.00	374.39	
	4.07	Lift shaft base - in situ concrete wall	0.00	422.41	
	5.00	Floor Finishes			
	5.01	Omit Armstrong Tapestry Plus	0.00	0.00	NOT used/claimed
		Add Sommer Century	0.00	0.00	Sommer Century
			0.00	0.00	Ditto to treads/risers
	6.01	Security system by Initial Shorrock	0.00	0.00	
	6.02	CCTV system by Initial Shorrock	0.00	0.00	
	6.03	No maintenance required	0.00	0.00	
	7.01	Power to illuminated nosings	0.00	0.00	Incl. elsewhere
	7.02	Omit south wing stair from quote	0.00	0.00	
	8.01	Lift - Add low level operating panel			
	8.02	Lift - Add text phone inside lift car			
21	1.00	Electrical Queries			
	1.01	DY16 - Omit lighting from gnd. flr.	-361.00	0.00	
	1.02	DY17 - CWS to provide cable to nosings	0.00	861.55	
	1.03	DY18 - disabled call button	0.00	61.86	
	1.04	DY18 - feed unit from switchgear	0.00	60.50	
	1.05	DY19 - A. I. issued			
		To summary	-361.00	2,495.71	

Figure A.2   (*Continued*)

Murribell Builders Ltd
City Centre Museum

Architect's Instructions

Interim Valuation No.5 - 28th July 2001

A.I. No.	Item	Description	Omit		Add	
21 (cont'd)	1.06	DY19 - A. I. issued				
	1.07	DY19 - A. I. issued				
	1.08	DY19 - CD to confirm				
	2.01	5th floor steel drawings - no comment	0.00		0.00	
22	1.01	Lower flat roof - recover blistered felt	0.00		852.50	
	2.01	Cables to intercom/text - normal cabling				
	3.01	New beam to south wing stair extension	0.00		0.00	see AI No.29/3.01
23		Electrical queries - DY20				
	1.01	Condenser in each shaft, etc.	0.00	50%	187.00	
	1.02	Item 2. Ex or MR 1500				
	1.03	Item 3. Data and telephone containment				
	1.04	Item 4. Hand drier mounted at 950mm				
		Electrical queries - DY23				
	1.05	Item 1. Number of ways to unit is 28	0.00		0.00	
	1.06	Item 2. Unit to be located at old shop	0.00		1,980.00	
	1.07	Item 3. Above system to basement to fifth floors				
	2.00	W.C./Shower Room - omit battery powered tap	0.00		0.00	
		- add mains powered tap	0.00		61.60	
24	1.01	Issue of drawings				
		G532/1L(--) 12e	0.00		0.00	
		G532/1L(--) 17c	0.00		0.00	
		G532/1A(31) 01d	0.00		0.00	
		G532/1A(66) 01b	0.00		0.00	
		G532/1S(31) 01c	0.00		0.00	
		G532/1S(31) 02b	0.00		0.00	
	2.01	45A switch to feed dark room unit				
	2.02	Alarm & Communication dwgs. - no comment	0.00		0.00	
25	1.01	Issue of drawings				
		G532/4/03 & bending schedule	0.00		0.00	
		G532/4/04 & bending schedule	0.00		2,427.97	
	1.02	2 copies of drawing schedule	0.00		0.00	
		To summary	0.00		5,509.07	

Figure A.2 *(Continued)*

Murribell Builders Ltd
City Centre Museum

Architect's Instructions

Interim Valuation No.5 - 28th July 2001

A.I. No.	Item	Description	Omit		Add	
25 (cont'd)	2.00	Electrical queries				
	2.01	Re AI 23/1.01 omit MICC wiring add S.W.A.				
	2.02	Re AI 22/2.01 omit MICC wiring add FP 200				
	2.03	Re AI 21/1.02 omit MICC wiring add FP 200				
26	1.01	Electrical 100kVA standby generator	0.00	20%	5,744.92	
	2.01	Ref. DY25 - fit separate braille label				
	2.02	Ref. DY26 1. D406 does not have electrical hardware 2. D303 has no electrical hardware D408 is incl. in Allgoods schematics 3. DG01,03,05,10,105,204 interface with alarm				
	2.03	Ref. PC3 - units supplied from basement etc				
27	1.01	Add Tarket Somer Standard to 4th flr. corridor				
	1.02	Two copies of dwg. G532/FLOOR-01, rev.B.				
	2.01	WC / Shower - height to push plate is 200mm				
	3.01	Refuge communication system - sign, etc.				
28	1.01	Dwg. issue G3882/G/03 rev.B				
	2.01	Illuminated nosings (12No.)as Bowey quote	0.00 0.00			Gradus stair nosings 2No. power units
	3.01	W.S. Controls - data cable works	0.00		679.36	
29	1.01	Dwg. issues G3882/F/03 rev.E and 04 rev.E	0.00 0.00	35% 5%	703.26 105.18	Additional luminaires Additional flashing beacons
	2.01	Position of text writer agreed				
	2.02	Timelock digital 24 hour 7 day programmable	0.00		48.94	
	3.00	Confirm COW's Directions				
	3.01	No.19 - extra beam for south west stair	0.00		136.75	
	3.02	No.20 - dwgs. G532/4/004 rev.B,03,04	0.00		0.00	
	3.03	No.21 - Omit 3/1/C5 & C6	-274.71		0.00	
	3.04	Plasterboard to extg. office (5th flr)	0.00		0.00	
	3.05	Extra wavy tail tie to lift shaft wall	0.00		167.66	
	3.06	No.22 - reposition dry riser valve				
		To summary	-274.71		7,586.07	

Figure A.2 (Continued)

Murribell Builders Ltd
City Centre Museum

Architect's Instructions

Interim Valuation No.5 - 28th July 2001

A.I. No.	Item	Description	Omit		Add	
29 (cont'd)	3.07	No.23 - trim out joists for hatches etc	0.00		32.29	
	3.08	4th flr - infill panel behind lift shaft wall	0.00		0.00	
	3.09	Blockwork to main entrance etc	0.00		0.00	
	3.10	Disabled WC - make good hole in floor	0.00		14.13	
	3.11	No.24 - WS Controls data cable in lift shaft	0.00		0.00	
	3.12	No.25 - dwg. G532/Floor 01 rev.A				
	3.13	Omit dormer to lift shaft etc	-476.01		0.00	
	3.14	No.26 - drill hole for CCTV system 150x650	0.00		203.50	
	3.15	Take off double door to extg. lift shaft etc	0.00		0.00	
	3.16	No.27 - fax from Dave Walton				
	3.17	DG01 form stud wall above blockwork	0.00		117.38	
	3.18	Continuous supply of concrete to walkway				
	3.19	No.28 - work to 600mm handrail at £26.95/m	0.00		0.00	Not including paint
30	1.01	Dwg. issues G3882/F/03 rev.F and 07 rev.B	0.00	50%	417.58	Additional luminaires 4th flr
			0.00		0.00	
	2.01	Electric door hardware as agreed at meeting	0.00		0.00	
	2.02	D105 and D204 have hold open devices	0.00		0.00	
	2.03	DG10 - salvage extg. hold open device?	0.00		0.00	
31	1.01	Air conditioning to extg. shaft as quote	0.00		25,196.81	
	2.01	New 48mm handrail to existing stair	0.00	0.00	0.00	Not including paint
32	1.01	Electrical shutdown CWS responsibility				
	2.01	Dwg. issue 1L(--)16 rev.C	0.00		0.00	
		To summary	-476.01		25,981.69	

Figure A.2 *(Continued)*

Murribell Builders Ltd
City Centre Museum

Architect's Instructions

Interim Valuation No.5 - 28th July 2001

	Omit	Add
Page 1	0.00	48,556.93
Page 2	0.00	3,098.21
Page 3	0.00	16,870.70
Page 4	0.00	14,372.53
Page 5	-5,935.99	22,264.67
Page 6	-953.48	7,572.59
Page 7	-83.30	9,690.67
Page 8	-9,929.36	26,547.06
Page 9	-361.00	2,495.71
Page 10	0.00	5,509.07
Page 11	-274.71	7,586.07
Page 12	-476.01	25,981.69
TOTAL	-18,013.85	190,545.90
		-18,013.85
		172,532.05

Figure A.2 (*Continued*)

Murribell Builders Ltd
City Centre Museum

Sundry Variations

Interim Valuation No.5 - 28th July 2001

S.V. No.	Description	Omit	Add	
1	Additional works to People's Gallery	0.00	0.00	A.I. No.17 / 1.01
2	Revised works to existing lift shaft	0.00	99.00	
3	Break out column bases in lift pit	0.00	0.00	A.I. No.17 / 2.01
4	Remove sprinkler heads to chem. store (CWS)	0.00	144.45	
5	Remove sprinkler heads to toilet/shower (CWS)	0.00	73.91	
6	100KVA standby generator	0.00	0.00	A.I. No.26 / 1.01
7	Replace glazing bars to roof glazing panels	0.00	769.16	
8	Bar reinforcement			
	Omit - 4/4/E9 - E11 (A.I. No.2/2.01)	-1,966.47	0.00	
	- Basement pit wall (A.I. No.2/2.01)	-350.00	0.00	
	- 5/3/E8 - E11	-195.53	0.00	
	Add - Lift pit base	0.00	501.38	
	- Plant room slab	0.00	253.98	
	- South block stairway extension	0.00	376.26	
9	Halfen channels to new lift shaft	0.00	1,511.47	
10	4No. powder coated roof vents (our fax 6/4/00)	0.00	0.00	A.I. No.14 / 3.02
11	Proctor Flooring - floorcovering variations :			
	1. Re-nail plywood to 4/5th floor (fax 12/7/00)	0.00 384m2	422.40	part only
	2. Latex screed to fill nail heads (fax 12/7/00)	0.00 384m2	464.64	part only
12	Cutting out steel beams to lift shaft areas	0.00	757.35	
13	CWS - electrical variations :			
	1. Hand drier to easy access WC	0.00 50%	150.17	
	2. Lift sub-main and isolator as dwg.3882/F/05B	0.00 10%	68.36	
	3. Sub-main to lift services DB (same dwg.)	0.00 10%	20.71	
	4. Additional DB's & cabling (same dwg.)	0.00 15%	175.39	
	5. 63A isolator for new invertor (C. Design inst.)	0.00	145.87	
	To summary	-2,512.00	5,934.50	

Figure A.3

Murribell Builders Ltd
City Centre Museum

Sundry Variations

Interim Valuation No.5 - 28th July 2001

S.V. No.	Description	Omit		Add
14	Roof coverings - approx. quants. adjustment			
	Omit 7 / 6 / H6	-5,850.00		0.00
	Add remeasure for slate coverings	0.00	153m2	4,475.25
	Omit 7 / 8 / H24	-41.16		0.00
	Add remeasure for lead slates	0.00	17No.	16.66
	Omit 7 / 9 / H30	-409.50		0.00
	Add remeasure for slates	0.00	17No.	165.75
	Add remove lead from timber gutters	0.00	20m	19.60
	Add remove lead from valleys	0.00	35m	34.30
	Add remove lead flashings	0.00	278m2	272.44
	Add extend lead outlets	0.00	15No	330.00
	Add lead saddles to dormer corners	0.00	14No.	123.20
15	Omit 3 / 3 / G2 ; Steel fittings (approx. quants)	-647.18		0.00
	Add remeasure	0.00	0.37t	556.88
	To summary	-6,947.84		5,994.08

Figure A.3 (Continued)

work equally well with handwritten submissions; the key for the contractor's quantity surveyor is to make sure that any amendments made by the client's quantity surveyor are accurately assessed and recorded with the amended valuation. Unilateral and sometimes global adjustments by the client's representative should be avoided at all cost. To overcome this situation if it does occur, the contractor's quantity surveyor should spend some time explaining the need for detail and accuracy with the valuation process, how this relates to the payment of subcontractors and how the valuation figures are required for the completion of the contractor's cost valuation comparison systems.

Appendix 1: A worked example   **139**

Murribell Builders Ltd
City Centre Museum

Sundry Variations - Summary

Interim Valuation No.10 - 30th January 2001

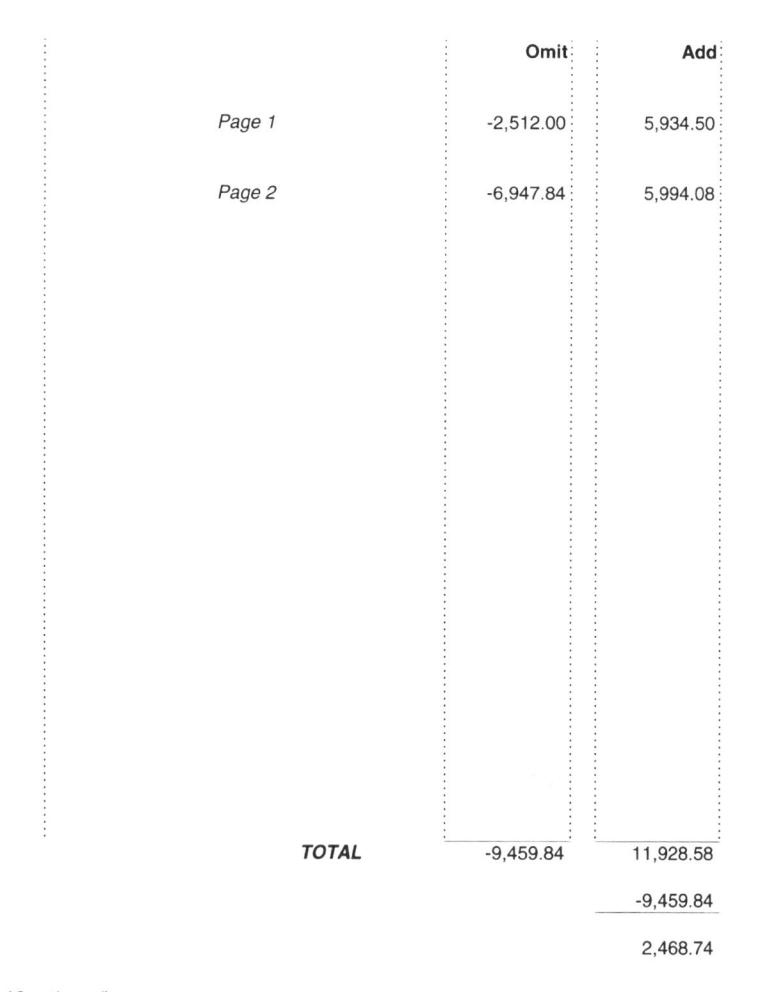

	Omit	Add
Page 1	-2,512.00	5,934.50
Page 2	-6,947.84	5,994.08
TOTAL	-9,459.84	11,928.58
		-9,459.84
		2,468.74

Figure A.3 (Continued)

Once final assessment has been made it is general practice for the client's quantity surveyor to issue an interim valuation certificate to the architect, copied to the contractor, for formal certification. In this example, however, no valuation certificate is issued. This is not an unusual situation when dealing with local authorities. It makes the monitoring of the architect's interim certificates and actual payments a little more difficult, but provided the contractor closely monitors the dates, this lack of a formal quantity surveyor's interim valuation should not

Murribell Builders Ltd

City Centre Museum

Materials on Site

Interim Valuation No.5 - 28th July 2001

item	quantity	unit	rate	total
Class B Engineering Bricks	450	No	0.13	58.50
Nails	2	Boxes	9.95	19.90
Rough Sawn Timber	0.19	m3	140.00	26.60
Dressed Softwood	0.86	m3	235.00	202.10
Mastercrete Cement	5	Bags	2.19	10.95
Plywood 6mm thick	29	Shts	5.64	163.56
Internal Doors	1	Item	2112.26	2112.26
Visqueen	1	Roll	29.00	29.00
Softwood Door Frames - single	1	No	9.90	9.90
Ironmongery	1	m2	6366.61	6366.61
50mm RW7 Insulation Slab	106	Say	6.44	682.64
Miscellaneous	1	Sum	150.00	150.00
Plasterers Material	1	Sum	5700.00	5700.00
Floorcovering Materials	1	Sum	23268.44	23268.44
Mechanical/Electrical Materials	1	Sum	26000.00	26000.00
Painters Materials	1	Sum	1000.00	1000.00
		To Valuation Summary		£65,800.46

Figure A.4

present any problems. In this example the 'date of issues' of the architect's interim certificate is 8 August 2000, with a valuation date of 1 August 2000 (Figure A.8). Although contractually these recorded dates are correct, the contractor's valuation date was 28 July 2000; in this instance, the weekend accounted for the difference, but the contractor needs to be aware of the critical nature of the certification dates and their relationship to payment dates and make the necessary representations should the contract timetable not be adhered to.

Once certification has been achieved the example indicates the next step as being notification or advice of payment due to the appropriate section of the contractor's office (Figure A.9). As explained in the main text, this is a very important document to the contractor and is used to monitor when payments should be made by respective clients. At this juncture the contractor's quantity surveyor should also make a diary note to check whether payment has been made. Since prompt payment is critical to all businesses, there can be no relaxation of the requirement to closely monitor and ensure timely payment of properly certified amounts.

With some projects employers will require a tax invoice to be issued by the contractor. The contractor's quantity surveyor needs to take particular care with this document, ensuring that it is accurately produced in all respects and particularly that all figures on the invoice are correct. Many payment systems installed

Appendix 1: A worked example **141**

Murribell Builders Ltd
City Centre Hotel
Interim Valuation No 5 - 28th July 2000

PQS amended valuation

Preliminaries

1/16 - Bond	%	100		444.00
1/63 - Management and Staff	%	87	39,156.00	34,065.72
1/64 - Site Accommodation	%	87	3,429.38	2,983.56
1/65 - Power - additional services	%	87	15,100.00	13,137.00
- Telephone installation	%	100		360.00
- Cleaning	%	20	800.00	160.00
1/66 - Hoist - erect/dismantle	%	50	1,860.00	930.00
- Hire	%	100		7,650.00
- Transport, concrete plant	%	100		910.00
1/67 - Scaffolding	%	95	44,753.00	42,553.35
Fans, fencing etc	%	87	17,636.00	15,343.32

4th & 5th Floor Alterations

3/1/C1 - 3/5/K4	%	100		24,458.25
3/5/K5-K7	%	75	10,733.74	8,050.31
K/8-K10	%	60	920.79	552.47
3/6/K11	%	60	43.95	26.37
K12	%	100		292.50
K13	%	85	6,316.18	5,368.75
K/15, K16	%	100		2,276.47
3/7/K17 - K23	%	95		7,459.74
3/8/K24 - K33	%	95		6,728.64
3/9/K34, K35	%	100		53.08
L1	No	5	458.15	2,290.75
L2	No	2	262.35	524.70
L3	No	2	139.03	278.06
L4	No	2	82.78	165.56
3/10/L7 - L10	%	87.5	426.18	372.91
L14-L17	%	67	360.46	174.51
3/11/L18	%	67	117.15	78.49
M1	%	100		29.30
3/12/M3	%	27	16,447.79	4,440.90
M6	%	50	6,106.24	3,053.12
3/13/M12	%	50	307.35	153.68
3/14/P1	%	85	1,002.54	852.16
P2,P3	%	50	458.64	229.32
P4	%	90	1,164.24	1,047.82
			c/fwd	187,494.81

Figure A.5

			b/fwd		187,494.81
4th & 5th Floor Alterations (continued)					
3/14/P5 - P7	%	100			8,778.00
3/15/P8-P11	%	50	11,874.92		5,937.46
3/16/P19-P27	%	100			1,015.88
3/17/P28-P32	%	100			3,856.74
Lift Shaft					
4/1/C1 - 4/7/L9	%	100			57,937.34
4/8/M1 - M5	%	100			4,366.55
4/9/P1, P2, P7	%	100			277.02
4/10/P8, P9	%	100			140.30
South Wing Staircase					
5/1/C1-5/4/G3	%	100			5,670.78
5/5/L1	%	1	359.32		359.32
L2 - L4	%	33	167.65		55.32
5/7/P5 - P7	%	100			334.90
People's Gallery					
6/1/C1 - F4	%	100			1,799.76
6/2/L1	%	100			273.63
6/3/L2 - M2	%	100			553.29
6/4/M4 - M7	%	100			775.18
6/8/P5 - P7	%	100			320.30
Roof					
7/1/C1 - C4, C9 - C11	%	100			3,989.89
7/2/C12-C22	%	100			9,566.22
7/3/G1 - G4	%	100			1,157.09
7/4/G5 - G16	%	100			1,814.42
7/5/G17, G21 - G25	%	100			1,011.65
7/6/H1 - H9	%	95			29,818.32
7/7/H10 - H19	%	100			5,959.49
7/8/H20 - H27	%	100			1,839.30
7/9/H28 - H34	%	100			2,170.35
7/10/H35 - H42	%	100			6,932.34
7/11/H43 - H47	%	100			10,174.42
			c/fwd		354,380.07

Figure A.5 (*Continued*)

			b/fwd		354,380.07
Roof (Continued)					
7/12/K7. L1	%		100		880.99
7/13/M1 - M6	%		100		233.51
7/14/P1 - P3	%		77	15,275.40	11,762.06
R1 - R5	%		100		322.07
7/15/R6 - R11	%		100		410.89
External Walkways					
8/1/C1 - C4	%		100		6,357.61
8/2/C5 - C11	%		100		3,676.17
8/3/C12 - E4	%		100		9,552.14
8/4/E5 - F1	%		100		839.20
8/5/G1 - G5	%		100		12,033.90
8/6/G6 - J4	%		100		9,768.13
8/7/J5	%		100		325.23
M1 - M3	%		50	933.43	466.72
P1, P2	%		100		1,940.50
Mechanical Installation					
9/1	%		75	94,463.00	70,847.25
Electrical Installation					
10/1	%		60.81	105,284.00	64,023.20
Lift Installation					
11/1 - see attached invoice. See AI No 2.01	%		100		86,092.50
Architect's Instructions					
As attached summary				sum	165,274.27
Sundry Variations					
As attached summary				sum	1,443.90
Material on Site					
As attached summary				sum	65,800.46
					866,430.77
		Less	Retention @ 5%		-44,513.19
					823,109.00
		Less	Previously certified No 4		-573,542.00
			Amount Due		**£249,567.00**

Figure A.5 (*Continued*)

Murribell Builders Ltd
City Centre Museum

Architect's Instructions

Interim Valuation No.5 - 28th July 2001   **PQS Amended**

A.I. No.	Item	Description	Omit	Add	
13 (cont'd)	3.00	Confirmation of fax			
	3.01	Lift pit - Omit visqueen tanking and korkpak	-83.30	0.00	
		Add servicised preprufe system	0.00	1,705.05	
		Electrical queries			
	3.02	Order 7.5kVA emergency lighting cubicle			
	3.03	Batteries as original spec. , i.e. not plante cells	0.00	5,566.14	50% increase capacity
	3.04	Do not provide earth leakage monitor to board			
	3.05	Upgrade pvc conduits to high impact	0.00		
	3.06	Omit contactor controlled lighting to 4th floor	0.00	0.00	see AI No.16/2.03
14	1.01	Finishes			
	2.01	Omit Rockwool RW3 , Add Rocksil RS60	0.00	0.00	
	3.00	Confirm COW's Directions			
	3.01	No.13 - Add 6mm cable at 4th floor	0.00	217.83	
	3.02	No.14 - Powder coating for roof vents	0.00	1,738.00	
	4.01	Omit electromagnetic locks to DG07 & D303	0.00	0.00	incl. AI No.12/1.01
15	1.01	Omit string skirting to stairs , Add sealant			
		Omit coved skirting , Add set in skirting 6/5/M12.13	66m 396.00		
		Add butt joint to be pointed with sealant	0.00 23m	27.60	
	1.02	Quote for Gradus Pathfinder nosings	0.00	0.00	
	2.02	Drawing issue - G532/FLOOR-01 rev.A	0.00	0.00	
	3.00	Electrical queries (ref. DY11 & DY12)			
	3.01	CWS will provide containment supplies	0.00		
	3.02	Darkroom - flush to walls , surface to ceiling			
		Dis. wc - flush to walls , surface to ceiling			
		Gflr cor. - flush to walls , surface to ceiling			
		Lift lobbies/corridors - new walls concealed			
		- existing walls surface			
	3.03	DY12 identical to DY11			
		To summary	-83.30	9,650.62	

Figure A.6

Murribell Builders Ltd
City Centre Museum

Architect's Instructions

Interim Valuation No.5 - 28th July 2001   **PQS Amended**

A.I. No.	Item	Description	Omit	Add	
20	1.01	Metal stair clarification/amendment	0.00	715.00	
	2.00	Handrails to west & south wing stairs	0.00	0.00	
	2.01	Quote for upgrading	0.00	0.00	
	- 2.07	existing and new handrail	0.00	0.00	
	3.01	Handrail to metal stair amended	0.00	0.00	
	4.00	Confirm COW's Directions - No.16			
	4.01	Vinyl sheet to stair is Armstrong Medintech	0.00	0.00	see AI No.6/4.00
	4.02	Vinyl to other areas is Armstrong Tapestry	0.00	0.00	NOT used/claimed
	4.03	Plywood fixings punch nailed	0.00	0.00	as BOQ's
		- No.17			
	4.04	Lift shaft base - break out column bases	0.00	0.00	see AI No.17/2.01
	4.05	Paint 3No. windows to water tower	0.00	0.00	
		- No.18			
	4.06	Drainage in lift shaft - replace c.i. pipe	0.00	367.79	
	4.07	Lift shaft base - in situ concrete wall	-498.18	0.00	
	5.00	Floor Finishes			
	5.01	Omit Armstong Tapestry Plus	0.00	0.00	NOT used/claimed
		Add Sommer Century	0.00	0.00	Sommer Century
			0.00	0.00	Ditto to treads/risers
	6.01	Security system by Initial Shorrock	0.00	0.00	
	6.02	CCTV system by Initial Shorrock	0.00	0.00	
	6.03	No maintenance required	0.00	0.00	
	7.01	Power to illuminated nosings	0.00	0.00	Incl. elsewhere
	7.02	Omit south wing stair from quote	0.00	0.00	
	8.01	Lift - Add low level operating panel			
	8.02	Lift - Add text phone inside lift car			
21	1.00	Electrical Queries			
	1.01	DY16 - Omit lighting from gnd. flr.	-361.00	0.00	
	1.02	DY17 - CWS to provide cable to nosings	0.00	861.55	
	1.03	DY18 - disabled call button	0.00	61.86	
	1.04	DY18 - feed unit from switchgear	0.00	60.50	
	1.05	DY19 - A. I. issued			
		To summary	-859.18	2,066.70	

Figure A.6 *(Continued)*

Murribell Builders Ltd
City Centre Museum

Architect's Instructions

Interim Valuation No.5 - 28th July 2001   PQS Amended

A.I. No.	Item	Description	Omit		Add	
25 (cont'd)	2.00	Electrical queries				
	2.01	Re AI 23/1.01 omit MICC wiring add S.W.A.				
	2.02	Re AI 22/2.01 omit MICC wiring add FP 200				
	2.03	Re AI 21/1.02 omit MICC wiring add FP 200				
26	1.01	Electrical 100kVA standby generator	0.00	20%	5,744.92	
	2.01	Ref. DY25 - fit separate braille label				
	2.02	Ref. DY26 1. D406 does not have electrical hardware 2. D303 has no electrical hardware D408 is incl. in Allgoods schematics 3. DG01,03,05,10,105,204 interface with alarm				
	2.03	Ref. PC3 - units supplied from basement etc				
27	1.01	Add Tarket Somer Standard to 4th flr. corridor				
	1.02	Two copies of dwg. G532/FLOOR-01 , rev.B.				
	2.01	WC / Shower - height to push plate is 200mm				
	3.01	Refuge communication system - sign , etc.				
28	1.01	Dwg. issue G3882/G/03 rev.B				
	2.01	Illuminated nosings (12No.)as Bowey quote	0.00 0.00			Gradus stair nosings 2No. power units
	3.01	W.S. Controls - data cable works	0.00		679.36	
29	1.01	Dwg. issues G3882/F/03 rev.E and 04 rev.E	0.00 0.00	35% 5%	703.26 105.18	Additional luminaires Additional flashing beacons
	2.01	Position of text writer agreed				
	2.02	Timelock digital 24 hour 7 day programmable	0.00		48.94	
	3.00	Confirm COW's Directions				
	3.01	No.19 - extra beam for south west stair	0.00		101.17	
	3.02	No.20 - dwgs. G532/4/004 rev.B,03,04	0.00		0.00	
	3.03	No.21 - Omit 3/1/C5 & C6	-274.71		0.00	
	3.04	Plasterboard to extg. office (5th flr)	0.00		0.00	
	3.05	Extra wavy tail tie to lift shaft wall	0.00		167.66	
	3.06	No.22 - reposition dry riser valve				
		To summary	-274.71		7,550.49	

Figure A.6 (*Continued*)

Murribell Builders Ltd
City Centre Museum

Architect's Instructions

Interim Valuation No.5 - 28th July 2001   **PQS Amended**

A.I. No.	Item	Description	Omit		Add	
29 (cont'd)	3.07	No.23 - trim out joists for hatches etc	0.00		32.29	
	3.08	4th flr - infill panel behind lift shaft wall	0.00		0.00	
	3.09	Blockwork to main entrance etc	0.00		0.00	
	3.10	Disabled WC - make good hole in floor	0.00		10.00	
	3.11	No.24 - WS Controls data cable in lift shaft	0.00		0.00	
	3.12	No.25 - dwg. G532/Floor 01 rev.A				
	3.13	Omit dormer to lift shaft etc	-476.01		0.00	
	3.14	No.26 - drill hole for CCTV system 150x650	0.00		70.68	
	3.15	Take off double door to extg. lift shaft etc	0.00		0.00	
	3.16	No.27 - fax from Dave Walton				
	3.17	DG01 form stud wall above blockwork	0.00		117.38	
	3.18	Continuous supply of concrete to walkway				
	3.19	No.28 - work to 600mm handrail at £26.95/m	0.00		0.00	Not including paint
30	1.01	Dwg. issues G3882/F/03 rev.F and 07 rev.B	0.00	50%	417.58	Additional luminaires 4th flr
			0.00		0.00	
	2.01	Electric door hardware as agreed at meeting	0.00		0.00	
	2.02	D105 and D204 have hold open devices	0.00		0.00	
	2.03	DG10 - salvage extg. hold open device?	0.00		0.00	
31	1.01	Air conditioning to extg. shaft as quote	0.00		20,000.00	
	2.01	New 48mm handrail to existing stair	0.00	0.00	0.00	Not including paint
32	1.01	Electrical shutdown CWS responsibility				
	2.01	Dwg. issue 1L(--)16 rev.C	0.00		0.00	
		To summary	-476.01		20,647.93	

Figure A.6   (*Continued*)

Murribell Builders Ltd
City Centre Museum
Architect's Instructions

Interim Valuation No.5 - 28th July 2001   PQS Amended

	Omit	Add
Page 1	0.00	48,556.93
Page 2	0.00	3,098.21
Page 3	0.00	16,870.70
Page 4	0.00	14,372.53
Page 5	-5,935.99	21,343.47
Page 6	-953.48	7,572.59
Page 7	-83.30	9,650.62
Page 8	-9,929.36	26,547.06
Page 9	-859.18	2,066.70
Page 10	0.00	5,509.07
Page 11	-274.71	7,550.49
Page 12	-476.01	20,647.93
TOTAL	-18,512.03	183,786.30
		-18,512.03
		165,274.27

Figure A.6 *(Continued)*

Murribell Builders Ltd
City Centre Museum

Sundry Variations

Interim Valuation No.5 - 28th July 2001   PQS Amended

S.V. No.	Description	Omit	Add	
1	Additional works to Peoples Gallery	0.00	0.00	A.I. No.17 / 1.01
2	Revised works to existing lift shaft	0.00	99.00	
3	Break out column bases in lift pit	0.00	0.00	A.I. No.17 / 2.01
4	Remove sprinkler heads to chem. store (CWS)	0.00	144.45	
5	Remove sprinkler heads to toilet/shower (CWS)	0.00	73.91	
6	100KVA standby generator	0.00	0.00	A.I. No.26 / 1.01
7	Replace glazing bars to roof glazing panels	0.00	769.16	
8	Bar reinforcement			
	Omit - 4/4/E9 - E11 (A.I. No.2/2.01)	-1,966.47	0.00	
	- Basement pit wall (A.I. No.2/2.01)	-350.00	0.00	
	- 5/3/E8 - E11	-195.53	0.00	
	Add - Lift pit base	0.00	501.38	
	- Plant room slab	0.00	253.98	
	- South block stairway extension	0.00	376.26	
9	Halfen channels to new lift shaft	0.00	1,511.47	
10	4No. powder coated roof vents (our fax 6/4/00)	0.00	0.00	A.I. No.14 / 3.02
11	Proctor Flooring - floorcovering variations :			
	1. Re-nail plywood to 4/5th floor (fax 12/7/00)	0.00 384m2	0.00	incl in B o Q
	2. Latex screed to fill nail heads (fax 12/7/00)	0.00 384m2	464.64	part only
12	Cutting out steel beams to lift shaft areas	0.00	757.35	
13	CWS - electrical variations :			
	1. Hand drier to easy access WC	0.00 50%	150.17	
	2. Lift sub-main and isolator as dwg.3882/F/05B	0.00 10%	68.36	
	3. Sub-main to lift services DB (same dwg.)	0.00 10%	20.71	
	4. Additional DB's & cabling (same dwg.)	0.00 15%	175.39	
	5. 63A isolator for new invertor (C. Design inst.)	0.00	145.87	
	To summary	-2,512.00	5,512.10	

Figure A.7

Murribell Builders Ltd
City Centre Museum

Sundry Variations

Interim Valuation No.5 - 28th July 2001

S.V. No.	Description	Omit		Add	
14	Roof coverings - approx. quants. adjustment				
	Omit 7 / 6 / H6	-5,850.00		0.00	
	Add remeasure for slate coverings	0.00	153m2	4,475.25	
	Omit 7 / 8 / H24	-41.16		0.00	
	Add remeasure for lead slates	0.00	17No.	16.66	
	Omit 7 / 9 / H30	-409.50		0.00	
	Add remeasure for slates	0.00	17No.	165.75	
	Add remove lead from timber gutters	0.00	20m	19.60	
	Add remove lead from valleys	0.00	35m	34.30	
	Add remove lead flashings	0.00	278m2	0.00	Incl 7/1.61
	Add extend lead outlets	0.00	15No	0.00	Incl 13/2.03
	Add lead saddles to dormer corners	0.00	14No.	123.20	
15	Omit 3 / 3 / G2 ; Steel fittings (approx. quants)	-647.18		0.00	
	Add remeasure	0.00	0.37t	556.88	
	To summary	-6,947.84		5,391.64	

Figure A.7 *(Continued)*

by employers, whether they are electronic or otherwise, will reject invoices containing what may appear to the contractor to be only typographical or peripheral errors, with delays in payment ensuing. In this example, the client does not require a formal invoice to be sent, as the architect's certificate acts as the contractor's documentation for tax purposes.

Once this initial paperwork has been completed, the contractor's quantity surveyor needs to move on to the next stage, i.e. the completion of information to be used in the production of the cost valuation comparison and subcontract payments. In general, the client's quantity surveyor will have no input into this stage of the valuation process.

Murribell Builders Ltd

City Centre Museum

Sundry Variations - Summary

Interim Valuation No.10 - 30th January 2001

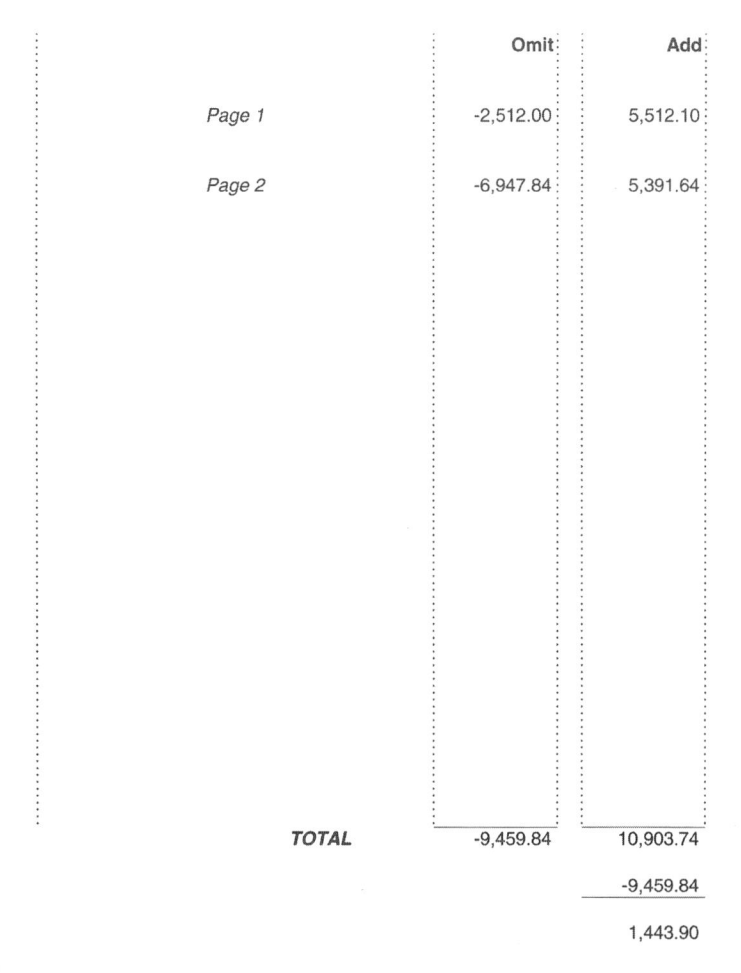

	Omit	Add
Page 1	-2,512.00	5,512.10
Page 2	-6,947.84	5,391.64
TOTAL	-9,459.84	10,903.74
		-9,459.84
		1,443.90

Figure A.7 *(Continued)*

The major area to be covered at this point is to establish the extent to which subcontract works have been included in the interim certificate. Again, there is a need for considerable accuracy in producing these figures. Chapter 5 covers the methodology behind the process and two individual subcontract elements are included here as a further example (Figures A.10 and A.11). Mechanical and electrical sections as well as roofing works are covered. The example indicates general valuation information as well as transferring precise details of works included in the

152 Commercial management in construction

Interim/Progress Payment

Issued by: address:	I V R Bone Head of Design Services City Museum Services Prehistoric Avenue Fossilton 3000000 BC
Employer: address:	City Museum Services Prehistoric Avenue Fossilton 3000000 BC
Contractor: address:	Murribell Builders Ltd High Street Sunniside England E1 2BB
Works: situated at:	City Centre Museum
Contract dated:	25/05/01

Serial no:	B 3447
Job reference:	G532
Certificate no:	5
Issue date:	08/08/00
Valuation date:	01/08/00
Contract sum:	£1,110,078.00

Contractor's copy

This Certificate for Interim/Progress Payment is issued under the terms of the above-mentioned Contract.

[1] Percentage is normally 95% except where Practical Completion has been achieved (97½%) or where some other percentage has been agreed by the parties.

[2] This item applies to IFC 84 only. Delete for MW 80 or if not relevant to this Certificate.

[1] A. Value of work executed and of materials and goods on site – (excluding items included in E below) .... £ 866,430.77

[1] B. Amount payable (95 % of A) .... £ 823,109.00

[2] C. Release of retained percentage on partial possession .... £

[2] D. Payment for goods and materials off site .... £

[2] E. Amount payable (or deductible) at 100% in accordance with IFC 84 clause 4·2·2 .... £

Sub-total £ 823,109.00

Less amounts previously certified .... £ 573,542.00

Net amount for payment .... £ 249,567.00

I/We hereby certify that the **amount for payment** by the Employer to the Contractor on this Certificate is (in words)

Two hundred and forty nine thousand five hundred and sixty seven pounds plus VAT.

All amounts are exclusive of VAT

To be signed by or for the issuer named above

Signed

[3] Relevant only if clause A1 of IC 81 Supplemental Conditions or clause B1 of MW 80 Supplementary Memorandum applies. Delete if not applicable.

[3] The Contractor has given notice that the rate of VAT chargeable on the supply of goods and services to which the Contract relates is 17.5 %

[3] 17.5 % of the amount certified above is .... £ 43,674.23

[3] Total of net amount and VAT amount (for information) .... £ 293,241.23

This is not a Tax Invoice

Figure A.8

Murribell Builders Ltd

Advice of Payment Due

Memorandum

To: Finance Department

From: DW   Date: 31-Jul-2000

Project	City Centre Museum
Project reference Nr	15681
Valuation Nr	5
Date on site	28-Jul-2000
PQS certificate date	
Date of Issue of Architect's Certificate	08-Aug-2000
Anticipated date payment due	09-Aug-2000

Gross amount of Architects Certificate	£866,430.77
Net payment due	£249,567.00
VAT	£43,674.23

Figure A.9

core of the main interim certificate. As indicated in Chapter 5, the requirement is to extract the exact quantities of subcontract works from the valuation and express them in monetary terms using both main contractor and subcontractor rates. In addition to the general measured elements of the valuation, the example shows how architect's instruction, sundry variation, materials on site and items of work carried out by the subcontractor on behalf of the contractor, but not included within the valuation, are dealt with.

As the example shows, these individual assessments are transferred to a subcontract liability summary sheet for further use within the cost valuation comparison and subcontract payment process. On the summary example sheet, subtrades have, for the sake of privacy of information, been included as trade descriptions only, whereas generally the subcontractors' names would have been used throughout.

In addition to the subcontract liabilities, before commencing work on the actual cost valuation comparison schedule, other areas of adjustment are required.

Preliminaries, for example, are included in the interim application using a simple scrolling principle through the bill of quantity items. By comparing the client's quantity surveyor's amended version of preliminary valuation (Figure A.5) to that of the contractors (Figure A.1), it can be seen that the client's quantity surveyor has added in an extra item for further scaffolding work (fans, fencing, etc.). This item was omitted in error by the contractor in the initial assessment. Despite such adjustments the contractor needs to be much more precise in considering preliminary works, and the example shows how a more accurate assessment is made to establish the true value of preliminary works before completion of the cost value comparison.

City Centre Museum

Contract No. 15681

Valuation No.5 - 28th July 2000
Subcontract Liabilities

Mechanical / Electrical Installation | **Murribell Builders Ltd** | **Subcontract**

Page	Bill Ref	BOQ's Quantity	Valuation Quantity	Rates	Total	Rates	Total
6 / 6	N1	1		2426.25	0.00	2883.00	0.00
	N3	1		1275.00	0.00	1707.87	0.00
	N4	1		2380.19	0.00	2412.25	0.00
6 / 7	N5	2		60.00	0.00	144.64	0.00
	N6	5		20.00	0.00	28.99	0.00
	N7	1		49.27	0.00	40.23	0.00
	N8	1		47.24	0.00	45.85	0.00
	N9	1		482.80	0.00	432.44	0.00
	N10	1		127.41	0.00	151.47	0.00
6 / 8	R1	1		3.34	0.00	5.46	0.00
	R2	3		1.80	0.00	1.71	0.00
	R3	13		9.31	0.00	10.14	0.00
	R4	1		7.39	0.00	6.74	0.00
	R5	2		8.17	0.00	9.55	0.00
	R6	2		10.01	0.00	13.52	0.00
6 / 9	R7	1		27.74	0.00	32.32	0.00
7 / 14	R1	10	10	19.63	196.30	36.73	367.30
	R2	2	2	19.09	38.18	20.74	41.48
	R3	1	1	33.98	33.98	30.01	30.01
	R4	1	1	19.63	19.63	36.73	36.73
	R5	1	1	33.98	33.98	24.71	24.71
7 / 15	R6	17	17	15.01	255.17	23.22	394.74
	R7	2	2	13.57	27.14	15.64	31.28
	R8	4	4	20.45	81.80	17.32	69.28
	R9	1	1	15.01	15.01	23.22	23.22
	R10	2	2	9.10	18.20	6.73	13.46
	R11	1	1	13.57	13.57	15.64	15.64
9 / 1	a	1	0.75	94463.00	70847.25	91000.00	68250.00
10 / 1	a	1	0.6081	105284.00	64023.20	98850.00	60110.69
					0.00		0.00
Murribell Builders Ltd					0.00		0.00
- meter		1	1	0.00	0.00	186.24	186.24
- socket		1	1	0.00	0.00	56.58	56.58
- lights		1	1	0.00	0.00	324.75	324.75
- spots		1	1	0.00	0.00	16.50	16.50
- trans.		1	1	0.00	0.00	38.90	38.90
- hoist		1	1	0.00	0.00	165.31	165.31
- micc cables		1	1	0.00	0.00	250.00	250.00
- micc cables		1		0.00	0.00	132.02	0.00
- joint box		1		0.00	0.00	120.00	0.00
- door holders		1		0.00	0.00	16.50	0.00
- strip meter		1		0.00	0.00	16.50	0.00
- switch		1		0.00	0.00	16.50	0.00
- strip power		1		0.00	0.00	115.51	0.00
- mods to bench		1		0.00	0.00	66.01	0.00
			C / Fwd.		135603.41		130446.82

Figure A.10

Appendix 1: A worked example   155

City Centre Museum

Contract No.   15681

Valuation No.5 - 28th July 2000
Subcontract Liabilities

		BOQ's	Valuation	Mechanical / Electrical Installation		Murribell Builders Ltd		Subcontract
Page	Bill Ref	Quantity	Quantity	Rates	Total	Rates	Total	
				B / Fwd.	135603.41	B / Fwd.	130446.82	
					0.00		0.00	
A.I. No.12	1.01	1	0.3	725.76	217.73	659.78	197.93	
	1.01	1	0.25	193.74	48.44	176.13	44.03	
	1.01	1		47.94	0.00	43.58	0.00	
	1.01	1		283.80	0.00	258.00	0.00	
					0.00		0.00	
A.I. No.13	3.03	1	1	5566.14	5566.14	5060.13	5060.13	
	3.04				0.00		0.00	
	3.05			0.00	0.00	400.16	0.00	
					0.00		0.00	
A.I. No.14	3.01	1	1	217.83	217.83	198.03	198.03	
					0.00		0.00	
A.I. No.15	3.01	1		3300.00	0.00	3000.00	0.00	
					0.00		0.00	
A.I. No.16	2.03	1	1	-619.46	-619.46	-619.46	-619.46	
					0.00		0.00	
A.I. No.17	1.01	1		5417.08	0.00	4908.78	0.00	
					0.00		0.00	
A.I. No.18	1.01	1	0.5	27385.60	13692.80	24896.00	12448.00	
					0.00		0.00	
A.I. No.20	6.00				0.00		0.00	
	Security	1		829.40	0.00	754.00	0.00	
	CCTV	1		1658.80	0.00	1508.00	0.00	
					0.00		0.00	
A.I. No.21	1.01	1	1	-361.00	-361.00	-361.00	-361.00	
	1.02	1	0.5	1723.10	861.55	1566.45	783.23	
	1.03	1	0.5	123.72	61.86	112.47	56.24	
	1.04	1	1	60.50	60.50	55.00	55.00	
					0.00		0.00	
A.I. No.23	1.01	1	0.5	374.00	187.00	340.00	170.00	
	1.05	1		421.00	0.00	471.50	0.00	
	1.06	1	1	1980.00	1980.00	1800.00	1800.00	
	2.00	1	0.7	88.00	61.60	80.00	56.00	
					0.00		0.00	
A.I. No.26	1.01	1	0.2	28724.62	5744.92	26113.29	5222.66	
					0.00		0.00	
A.I. No.29	1.01	1	0.35	2009.30	703.26	1826.64	639.32	
	1.01	1	0.05	2103.65	105.18	1912.41	95.62	
	2.02	1	1	48.94	48.94	44.49	44.49	
					0.00		0.00	
A.I. No.30	1.01	1	0.5	835.15	417.58	759.23	379.62	
	1.01	1		878.61	0.00	798.74	0.00	
					0.00		0.00	
A.I. No.31	1.01 - M	1	0.81	21401.33	17289.84	19455.75	15718.03	
	- E	1	1	867.33	867.33	788.48	788.48	
					0.00		0.00	
A.I. No.37	1.02	1		72.61	0.00	66.01	0.00	
	1.03	1		391.24	0.00	355.67	0.00	
					0.00		0.00	
A.I. No.39	3.02	1		915.75	0.00	832.50	0.00	
					0.00		0.00	
A.I. No.42	1.35	1		318.63	0.00	289.66	0.00	
	1.49	1		72.61	0.00	66.01	0.00	
					0.00		0.00	
A.I. No.43	1.01	1		41.49	0.00	37.72	0.00	
					0.00		0.00	
					0.00		0.00	
				C / Fwd.	182755.44		173223.16	

Figure A.10   (Continued)

City Centre Museum

Contract No. 15681

Valuation No.5 - 28th July 2000
Subcontract Liabilities

Mechanical / Electrical Installation **Murribell Builders Ltd** **Subcontract**

Page	Bill Ref	BOQ's Quantity	Valuation Quantity	Rates	Total	Rates	Total	
				B / Fwd.	182755.44	B / Fwd.	173223.16	
					0.00		0.00	
S.V. No.4		1	1	144.45	144.45	131.32	131.32	
					0.00		0.00	
S.V. No.5		1	1	73.91	73.91	67.19	67.19	Part only
S.V. No.6		1	0.02	28724.62	574.49	26113.29	522.27	
S.V. No.13.1		1	0.5	300.33	150.17	273.03	136.52	
S.V. No.13.2		1	0.1	683.64	68.36	621.49	62.15	
S.V. No.13.3		1	0.1	207.14	20.71	188.31	18.83	
S.V. No.13.4		1	0.15	1169.27	175.39	1062.97	159.45	
S.V. No.13.5		1	1	145.88	145.88	132.61	132.61	
S.V. No.13.6		1		61.72	0.00	56.11	0.00	
S.V. No.13.7		1		830.39	0.00	754.90	0.00	
S.V. No.13.8		1		174.78	0.00	158.89	0.00	
S.V. No.13.9		1		550.00	0.00	500.00	0.00	
S.V. No.13.10		1		282.67	0.00	256.97	0.00	
S.V. No.13.11		1		2420.00	0.00	2200.00	0.00	
S.V. No.13.12		1		0.00	0.00	0.00	0.00	see AI No. 37/1.0
S.V. No.13.13		1		143.00	0.00	130.00	0.00	
S.V. No.13.14		1		55.00	0.00	50.00	0.00	
S.V. No.13.15		1		84.70	0.00	77.00	0.00	
S.V. No.13.16		1		88.00	0.00	80.00	0.00	
S.V. No.13.17		1		220.00	0.00	200.00	0.00	
S.V. No.13.18		1		22.00	0.00	20.00	0.00	
S.V. No.13.19		1		11.41	0.00	10.37	0.00	
S.V. No.33		1		0.00	0.00	396.06	0.00	
					0.00		0.00	
					0.00		0.00	
					0.00		0.00	
					0.00		0.00	
					0.00		0.00	
M.O.S.			1	26200.00	26200.00	27700.00	27700.00	
					0.00		0.00	
					0.00		0.00	
					0.00		0.00	
					0.00		0.00	
					0.00		0.00	
					0.00		0.00	
			Total		210308.80		202153.49	

Figure A.10 (Continued)

Despite the best endeavours to arrive at an accurate valuation assessment, it will generally be necessary to make adjustments to the valuation. The example shows the detail necessary to produce an accurate value to set against the project cost. The adjustment in this case shows, among other items, the overrecovery of

Appendix 1: A worked example 157

City Centre Museum

Contract No. 15681

Valuation No.5 - 28th July 2000
Subcontract Liabilities

			Roofing Works		Murribell Builders Ltd		Subcontract	
	Page	Bill Ref	BOQ's Quantity	Valuation Quantity	Rates	Total	Rates	Total
	7 / 1	C2	68	68	2.44	165.92	2.50	170.00
		C3	29	29	9.75	282.75	10.00	290.00
		C4	1146	1146	incl.	0.00	incl.	0.00
		C5	1146		incl.	0.00	incl.	0.00
A.Q.		C6	100		1.95	0.00	2.00	0.00
A.Q.		C7	25		1.95	0.00	2.00	0.00
A.Q.		C8	10		1.95	0.00	2.00	0.00
		C9	30	30	0.98	29.40	1.00	30.00
		C10	148	148	0.98	145.04	1.00	148.00
		C11	1146	1146	2.93	3357.78	3.00	3438.00
	7 / 2	C16	40	40	9.75	390.00	10.00	400.00
	7 / 6	H4	1131	1074.45	19.50	20951.78	20.00	21489.00
		H5	19	18.05	19.50	351.98	20.00	361.00
A.Q.		H6	200	190	29.25	5557.50	30.00	5700.00
		H7	20	19	1.95	37.05	2.00	38.00
		H8	32	30.4	1.95	59.28	2.00	60.80
		H9	106	100.7	1.95	196.37	2.00	201.40
	7 / 7	H10	45	45	9.75	438.75	10.00	450.00
		H11	183	183	9.75	1784.25	10.00	1830.00
		H12	16	16	9.75	156.00	10.00	160.00
		H13	39	39	5.85	228.15	6.00	234.00
		H14	58	58	11.70	678.60	12.00	696.00
		H15	31	31	58.50	1813.50	60.00	1860.00
		H16	1	1	11.70	11.70	12.00	12.00
		H17	11	11	14.63	160.93	15.00	165.00
		H18	37	37	14.63	541.31	15.00	555.00
		H19	10	10	14.63	146.30	15.00	150.00
	7 / 8	H20	8	8	1.46	11.68	1.50	12.00
		H21	4	4	1.46	5.84	1.50	6.00
		H22	801	801	0.49	392.49	0.50	400.50
		H23	126	126	0.49	61.74	0.50	63.00
A.Q.		H24	42	42	0.98	41.16	1.00	42.00
		H25	801	801	1.17	937.17	1.20	961.20
		H26	126	126	1.17	147.42	1.20	151.20
		H27	31	31	7.80	241.80	8.00	248.00
	7 / 9	H28	93	93	11.70	1088.10	12.00	1116.00
		H29	8	8	11.70	93.60	12.00	96.00
A.Q.		H30	42	42	9.75	409.50	10.00	420.00
				C / Fwd.		40914.83		41954.10

Figure A.11

City Centre Museum

Contract No. 15681

Valuation No.5 - 28th July 2000
Subcontract Liabilities

			Roofing Works			Murribell Builders Ltd		Subcontract	
				BOQ's	Valuation				
	Page	Bill Ref		Quantity	Quantity	Rates	Total	Rates	Total
						B / Fwd.	40914.83	B / Fwd.	41954.10
	7 / 9	H31		7	7	11.70	81.90	12.00	84.00
		H32		7	7	11.70	81.90	12.00	84.00
		H33		13	13	11.70	152.10	12.00	156.00
		H34		27	27	9.75	263.25	10.00	270.00
	7 / 10	H35		7	7	9.75	68.25	10.00	70.00
		H36		22	22	9.75	214.50	10.00	220.00
		H37		44	44	9.75	429.00	10.00	440.00
		H38		148	148	29.25	4329.00	30.00	4440.00
		H39		10	10	24.38	243.80	25.00	250.00
		H40		35	35	29.25	1023.75	30.00	1050.00
		H41		20	20	29.25	585.00	30.00	600.00
		H42		8	8	4.88	39.04	5.00	40.00
							0.00		0.00
	7 / 2	C17		31	31	3.29	101.99	2.00	62.00
		C18		13	13	2.77	36.01	3.00	39.00
		C19		116	116	1.89	219.24	1.00	116.00
		C20		73	73	2.35	171.55	1.00	73.00
		C21		143	143	3.89	556.27	3.90	557.70
							0.00		0.00
	A.I. No.10	1.11		1	1	4415.40	4415.40	4014.00	4014.00
							0.00		0.00
	A.I. No.10	1.18		1	1	-223.08	-223.08	-223.08	-223.08
							0.00		0.00
	A.I. No.11	1.03		1	1	356.40	356.40	324.00	324.00
							0.00		0.00
7/10/H39	A.I. No.13	2.03		1	1	121.90	121.90	125.00	125.00
							0.00		0.00
	A.I. No.22	1.01		1	1	852.50	852.50	775.00	775.00
							0.00		0.00
	S.V. No.14	roof covering		1	1	-5850.00	-5850.00	-6000.00	-6000.00
		remeasure		1	1	4475.25	4475.25	4590.00	4590.00
		lead slates		1	1	-41.16	-41.16	-42.00	-42.00
		remeasure		1	1	16.66	16.66	17.00	17.00
		7 / 9 / H30		1	1	-409.50	-409.50	-420.00	-420.00
		remeasure		1	1	165.75	165.75	170.00	170.00
		lead from gut		1	1	19.60	19.60	20.00	20.00
		lead from val		1	1	34.30	34.30	35.00	35.00
		remove flash		0	0	272.44	0.00	278.00	0.00
		lift shaft lead		1		97.50	0.00	100.00	0.00
		lead saddles		1	1	123.20	123.20	112.00	112.00
		lead aprons		1		234.00	0.00	240.00	0.00
		cuts to slates		1		163.80	0.00	168.00	0.00
		lead to vent b		1		0.00	0.00	240.00	0.00
		snowboards		1		193.60	0.00	176.00	0.00
	Misc.			1	1	0.00	0.00	85.00	85.00
					Total		53568.60		54087.72

Figure A.11 (Continued)

Appendix 1: A worked example 159

Murribell Builders Ltd
ANALYSIS OF SUBCONTRACT PROFIT (Sheet 1 of 2)

| Contract | City Centre Museum | | | | Val'n No. | 5 | | | | |
| Contract no | 15681 | | | | P/Ending | 28/07/00 | | | | |

| Subcontractor | Order Value | MURRIBELL BUILDERS LTD | | | SUBCONTRACTOR | | | Ddt Disc. | | Total Liability | Profit |
		Measured Work	Stock on Site	Total Value	Measured Work	Stock on Site	Sub Total	%	Disc Allnce		
Lift Installation	87300.00	93,248		93,248	93,805		93,805	2.5	2,345	91,460	1,788
Mech & Elect Installn	196350.00	210,309		210,309	202,153		202,153	2.5	5,054	197,099	13,210
Roof Tiling	51393.00	53,569		53,569	54,088		54,088	3.5	1,893	52,195	1,374
Stainless Steel Rfg	10174.43	7,129		7,129	7,849		7,849	2.5	196	7,653	-524
Scaffolding	25521.00	30,572		30,572	29,481		29,481	0	0	29,481	1,091
Wall Tiling	369.15	0		0	0		0	2.5	0	0	0
Mastic Asphalt	3051.00	2,975		2,975	3,051		3,051	2.5	76	2,975	0
Suspended Clgs	1,253.20	0		0	0		0	2.5	0	0	0
Plasterwork	72,438.90	72,242		72,242	74,074		74,074	6	4,444	69,630	2,612
Decorating	11,514.27	3,907		3,907	4,011		4,011	6	241	3,770	137
Floor Coverings	17,717.06	31,696		31,696	31,643		31,643	2.5	791	30,852	844
Door Entry System	1,000.00	1,100		1,100	1,000		1,000	0	0	1,000	100
Handrails	15,044.00	34,108		34,108	34,108		34,108	2.5	853	33,255	853
Sundry Metalwork	2,585.00	5,240		5,240	3,924		3,924	2.5	98	3,826	1,414
Bird Prevention	17,265.00	11,395		11,395	10,359		10,359	2.5	259	10,100	1,295
Stone Restoration	7,940.00	13,519		13,519	12,290		12,290	2.5	307	11,983	1,536
Glazing	631.00	1,901		1,901	1,728		1,728	2.5	43	1,685	216
Electric Controls	Agreed Rates	679		679	618		618	0	0	618	61
				0			0		0	0	0
				0			0		0	0	0
				0			0		0	0	0
C / Fwd.		573,589	0	573,589	564,182	0	564,182		16,601	547,581	26,008

Figure A.12

Murribell Builders Ltd
ANALYSIS OF SUBCONTRACT PROFIT (Sheet 2 of 2)

Contract: City Centre Museum
Contract no: 15681
Val'n No.: 5
P/Ending: 28/07/00

| Subcontractor | Order Value | MURRIBELL BUILDERS LTD ||| SUBCONTRACTOR |||| Ddt Disc. || Total Liability | Profit |
		Measured Work	Stock on Site	Total Value	Measured Work	Stock on Site	Sub Total	%	Disc Allnce		
B / Fwd.		573,589		573,589	564,182		564,182		16,601	547,581	26,008
Aerials				0			0	0	0	0	0
Sundry Metalwork	Agreed Rates	200		200	200		200	2.5	0	200	0
General Cleaning	300.00	0		0	0		0	0	0	0	0
Core Drilling		389		389	370		370	0	0	370	19
				0			0	0	0	0	0
				0			0	0	0	0	0
				0			0		0	0	0
				0			0		0	0	0
				0			0		0	0	0
				0			0		0	0	0
				0			0		0	0	0
				0			0		0	0	0
				0			0		0	0	0
				0			0		0	0	0
				0			0		0	0	0
				0			0		0	0	0
				0			0		0	0	0
				0			0		0	0	0
TOTALS		574,178	0	574,178	564,752	0	564,752		16,601	548,151	26,027

Figure A.12 (Continued)

Murribell Builders Ltd
Internal Preliminaries

Contract No: 15681

Project: City Centre Museum

Valuation No: 5

Original Total	Preliminary Element	Total Weeks	Amount per week	Number of weeks	Total
13112.00	Site Supervision	22	596.00	19	11324.00
1750.00	Ganger	20	87.50	17	1487.50
8010.00	Barrow hoist	18	445.00	17	7565.00
300.00	Remove/Reinstate windows	Sum	300.00	0.5	150.00
1560.00	Mini hoist	10	156.00	5	780.00
200.00	Mixer Half bag	10	20.00	10	200.00
400.00	Stihl saw - purchase	Sum	400.00	0.85	340.00
210.00	- blades	21	10.00	18	180.00
210.00	- fuel	21	10.00	18	180.00
255.00	Rip snorter	15	17.00	10	170.00
160.00	Barrows	8No	20.00	8	160.00
880.00	Pace breaker/Kango	22	40.00	21	840.00
90.00	Whacker	2	45.00		0.00
560.00	Concrete vibrator	16	35.00	16	560.00
612.04	Drills (9No)	22	27.82	20	556.40
710.00	Transport plant to site	Sum	710.00	0.5	355.00
35345.00	Scaffolding - main	Sum	35345.00	0.75	26508.75
2000.00	- adaptions	Sum	2000.00		0.00
500.00	- aluminium towers	Sum	500.00	1	500.00
100.00	- trestles	Sum	100.00	1	100.00
1000.00	- allowance	Sum	1000.00		0.00
1848.00	- hoarding to base	Sum	1848.00	0.75	1386.00
400.00	- window removal	Sum	400.00	1	400.00
600.00	- window refit	Sum	600.00		0.00
50.00	Site office - transport	Sum	50.00	1	50.00
50.00	- erect/dismantle	Sum	50.00	1	50.00
462.00	- hire	21	22.00	18	396.00
210.00	- heat and lighting	21	10.00	18	180.00
50.00	Client's office - transport	Sum	50.00	1	50.00
50.00	- erect/dismantle	Sum	50.00	1	50.00
462.00	- hire	21	22.00	18	396.00
210.00	- heat and lighting	21	10.00	18	180.00
50.00	Canteen - transport	Sum	50.00	1	50.00
50.00	- erect/dismantle	Sum	50.00	1	50.00
462.00	- hire	21	22.00	18	396.00
210.00	- heat and lighting	21	10.00	18	180.00
50.00	Toilets - transport	Sum	50.00	1	50.00
50.00	- erect/dismantle	Sum	50.00	1	50.00
693.00	- hire	21	33.00	18	594.00
500.00	- connection	Sum	500.00		0.00
400.00	Telephone (2No) - installation	Sum	400.00	1	400.00
160.00	- rental	Sum	160.00	0.5	80.00
630.00	- calls	21	30.00	18	540.00
50.00	Fax paper	Sum	50.00	1	50.00
500.00	Temporary water - connection	Sum	500.00		0.00
250.00	- site service	Sum	250.00	0.85	212.50
1650.00	- charges	Sum	1650.00	1	1650.00
1500.03	Temporary electricity - connection	21	71.43	18	1285.74
749.91	- site service	21	35.71	18	642.78
199.92	- charges	21	9.52	18	171.36
700.00	Clear site as work proceeds	10	70.00	8	560.00
800.00	Final clean	Sum	800.00		0.00
100.00	Protect / Clean roads	Sum	100.00	0.85	85.00
13470.60	Internal temporary hoardings	471m2	28.60	410	11726.00
1665.00	Access doors to hoardings	9No.	185.00	8	1480.00
7350.00	Service gangs	21	350.00	18	6300.00
1375.08	Importation	21	65.48	18	1178.64
8298.99	Non-productive works	21	395.19	18	7113.42
1000.00	Special insurances	Sum	1000.00	1	1000.00
444.00	Performance bond	Sum	444.00	1	444.00
200.00	Winter working	Sum	200.00	0.85	170.00
200.00	Concrete testing	Sum	200.00	0.85	170.00
450.00	Sign board - supply and erect	Sum	450.00	1	450.00
50.00	- dismantle	Sum	50.00		0.00
1000.00	Glass breakage	Sum	1000.00		0.00
700.00	Protection to existing fittings	Sum	700.00	1	700.00
1270.00	Attendance on subs - conc drilling	Sum	1270.00	1	1270.00
3000.00	- general	Sum	3000.00	0.5	1500.00
800.00	Small tools	Sum	800.00	1	800.00
123384.57	Preliminary sub total				96444.09
149.00	Firm price allowance - staff	Sum	149.00	0.5	74.50
1021.00	- labour	Sum	1021.00	0.5	510.50
989.00	- material	Sum	989.00	0.5	494.50
500.00	- subcontract	Sum	500.00		0.00
200.00	Considerate constructors - fee	Sum	200.00	1	200.00
126243.57	Total Preliminaries				97723.59
6255.00	Late tender adjustments	Sum	6255.00	0.75	4691.25
132498.57	Total Internal Preliminaries				102414.84
	External Preliminaries from Valuation				118536.95
	Over / Under recovery of Preliminaries				(£16,122.11)

Figure A.13

City Centre Museum

Contract Number 15681

Interim Valuation No 5 - 28 July 2000

Valuation Adjustment		
Preliminaries		
External prelims less internal prelims		
118536.95 - 102,414.86		-16,122.09
Measured Work		
5/3/F1 BoQ's less measure 36 - 10 = 26 x 31.20 = 811.20		
F2 BoQ's less measure 8 - 8 = - -		
F3 BoQ's less measure 8 - 4 = 4 x 83.01 - 335.64		-1,146.84
7/6/H1 - H3 Adjusted by PQS to 95%, change back to 100%		140.23
H6 - Adjusted by PQS to 95%, change back to 100%		292.50
Doors in general as attached schedule		-9,248.40
Architect's Instructions		
10/1.17 Brickwork to inner wall - not done.		-953.50
13/2/11 New door - ironmongery and decoration still to do		-213.58
Sundry Variations		
7 - Replace glazing bars - damaged by Bowey?		-769.16
9 - Halfen channels - should have been in Kone quote		-1,511.47
Material on Site		
Painters material - not in subcontract liabilities		-1,000.00
Doors (2,112.26 - 779.95)		-1,332.31
Ironmongery (6,366.612 - 5,366.61)		-1,000.00
Supalux	sum	-2,119.00
Overvalue Transferred to Cost Value Comparison		**-£34,982.62**

Figure A.14

preliminary elements, general adjustments of measured work sections and adjustments to materials on site (Figures A.13–A.15).

The completion of the cost valuation comparison is as described in Chapter 5. The example shows how the various schedules described briefly above are inserted into their respective places on the comparison schedule (Figure A.1). It can be seen how the gross certified amount of £866,431.00 from the architect's

Contract Number 15681

Interim Valuation No 5 - 28 July 2000

Valuation Adjustment				
Doors				
Claimed in valuation	3/9/L1	5 No	458.15	2,290.75
	L2	2 No	262.35	524.70
	L3	2 No	139.03	278.06
	L4	2 No	82.78	165.56
	5/5/L1	1 No	359.32	359.32
	6/2/L1	3 No	91.21	273.63
AI No 2.01	4/6/L1	30 No	458.15	13,744.50
	4/7/L2	1 No	359.32	359.32
	L3	6 No	142.73	856.38
		52 No		18,852.22
		of 68 No in		
		total in BoQ's		
From door schedule G532 15(31)01 Revision C				
55 No doors in total, 10 No still to fix (material on site)				
52 - 45 =		7 No	458.15	3,207.05
Half annular door openings priced on but not used, standard rectangular				
opening used.				
437.82 - 265.21		35 No	172.61	6,041.35
Total in Summary				£9,248.40

Figure A.14 *(Continued)*

interim valuation certificate no. 5 is inserted into the appropriate box and how that figure becomes the starting figure of the cost value comparison. The value is then adjusted to cater for the overvaluation considered necessary by the contractor's quantity surveyor before making the deductions for subcontract liabilities

CITY CENTRE MUSEUM

Interim Valuation No 5 - 28 July 2000

Other Costs

M Kelly (Labour only steelfixing)	£969.00
Temporary Water Charge	£1,650.00
Temporary Electric Charge	
Final Clean	
Special Insurance Cover	£1,000.00
Bond Fee	£1,050.00
Considerate Constructor Fee	£200.00
Van Hire (?)	£38.45
Transfer to Cost Value Comparison	**£6,907.45**

Figure A.15

and actual costs incurred by the contractor. Within the compilation of these figures the example shows the schedule of 'other cost', i.e. those elements of cost that cannot be attributed to any other cost head section.

Many elements can be analysed from the cost value comparison (Figure A.16). In the example shown the assessment of time indicates that the project is forecast to overrun by five weeks on contract time and six weeks over the originally priced preliminary period. The schedule also indicates that an extension of time has been applied for to cover this overrun period, although no award has yet been made. The comments column confirms that no allowance has been made to adjust the preliminaries and also that there is no allowance in the figures to cater for the possible deduction of liquidated and ascertained damages. The decision not to make any allowance for any of the above criteria must be a team decision and must also prompt an action plan to expedite satisfactory conclusion of the overrun situation as quickly as possible. It would be of benefit to the management team if such action plans were also included in the comments box on the cost value comparison.

The cost value comparison indicates that all elements of the initial tender allowances have been exceeded: labour, plant, material and subcontract figures all exceed their respective initial tender assessments. In total, there is an overspend of £119,688.00 on current figures. However, if a comparison is drawn between this calculated figure and the assessment of variations issued to date, it can be concluded that the project has considerably increased in value, which could account for the additional spend throughout the resource costing. Whatever the situation, the figures indicated by or deduced from the cost value comparison will prompt further discussion among team members.

MURRIBELL BUILDERS LTD

Profit Appraisal No	5	at	28/07/00			

Certificate No	5	Contract	City Centre Museum	C.Manager	WIC	
Gross Value	866431	Location	Any town	Q.Surveyor	DW	
Valuation Date	28/07/00	Contract Nc	15681	Agent	ES	
Retention	43,322	Estimator	MB	Bonus	DK	
					Contract	Prelims
TENDER SUM	1110078	Contract Commencement Date			06/03/00	06/03/00
Less Dir.pymts.		Contract Completion Date			06/08/00	30/07/00
P.C.Adjustment	-334220	Contract Weeks Ahead/(behind)			-5	-6
Variations	270000	Revised Completion Date			10/09/00	10/09/00
					Contract	Prelims
Projected F/Acc...........£	1045858	Ext.App.For	5	Weeks	21/22	21/21
Less Adjusted Value	831448	Ext.Granted		Weeks	95.45%	100.00%
Outstanding Value........£	214,410	Ant.F/Awd		Weeks		
% Complete	79%			GROSS CERT VALUE		866431
				Sundry Invoices		

Tender	Budget	Forecast	Percentage			
0	53819	38500	3.82%			866431
				Val. Adjustment		-34,983
				GROSS ADJ. VAL.		831448
			Tender			
			0	Nom. Sub Contractor		
			0	Nom. Supplier		
			489805	Direct Sub Contractor	574178	-574,178
				GROSS VALUE OF OWN WORK		257270
				Snagging and Defects	1.5%	-3,859

Increase		Tender		Total	253,411
30,206	Labour	116962	Labour	132861	
13,629	Materials	79084	Materials	88754	
0	Plant Internal	0	Plant Internal		
5,080	External	17125	External	26871	
0	Haulage	0	Haulage		
38	Other Costs	3094	Other Costs	6907	255393
48,953					
189,090	S/c	Tender	Period	Accum	GROSS MARGIN
238,043		Percentage	Percentage	Percentage	Own Works	-1,982
5,327			10.88%	-0.78%	Nom. Sub Contractor	
0					Nom. Supplier	
0					Direct Sub Contractor	26,027
8,962			4.74%	4.53%		
14,289		0.00%	6.00%	2.99%	TOTAL TO DATE	**24,045**

COMMENTS	INITIALS
No allowance made with comparison for anticipated overrun either with regard to liquidated and ascertained damages or paid extension of time this despite current overrun of 5 weeks on contract time	
	ISSUED

Figure A.16

In terms of profitability, the cost value comparison denotes a disappointing situation, with the contractor's own element of work indicating a minor loss situation. However, the contractor can take some comfort from the fact that the profit in the costing period shows an improvement of £5327.00 on the contractor's element and the overall profit level is still indicating an enhancement of final profit anticipated from the project.

Appendix 2

Cost Value Comparison: An Easy Guide

Murribell Builders Ltd

Profit Appraisal No	-	at	-		

Certificate No	-	Contract	-	C.Manager	-
Gross Value	-	Location	-	Q.Surveyor	-
Valuation Date	-	Contract No	-	Agent	-
Retention	-	Estimator	-	Bonus	-

			Contract	Prelims
TENDER SUM	-	Contract Commencement Date	-	-
Less Dir.pymts.	-	Contract Completion Date	-	-
P.C.Adjustment	-	Contract Weeks Ahead/(behind)	-	-
Variations	-	Revised Completion Date		
			Contract	Prelims
Projected F/Acc............£	-	Ext.App.For - Weeks	-	-
Less Adjusted Value	-	Ext.Granted - Weeks	-	-
Outstanding Value........£	-	Ant.F/Awd - Weeks	-	-
% Complete	-	GROSS CERT VALUE		-
		Sundry Invoices		-

Tender	Budget	Forecast	Percentage		
-	-	-	-	Val. Adjustment	-
				GROSS ADJ. VAL.	-
			Tender		
			-	Nom. Sub Contractor	-
			-	Nom. Supplier	-
			-	Direct Sub Contractor	-
				GROSS VALUE OF OWN WORK	-
				Snagging and Defects	-

Increase			Tender		Total	
-	Labour		-	Labour	-	
-	Materials		-	Materials	-	
-	Plant	Internal	-	Plant	Internal	-
		External	-		External	
-	Haulage		-	Haulage	-	
-	Other Costs		-	Other Costs	-	-

		Tender	Period	Accum		
-	S/c	Percentage	Percentage	Percentage	GROSS MARGIN	
-	Total	-	-	-	Own Works	-
					Nom. Sub Contractor	
					Nom. Supplier	
-		-	-	-	Direct Sub Contractor	-
-		-	-	-	TOTAL TO DATE	-

COMMENTS	INITIALS
	ISSUED

168 Commercial management in construction

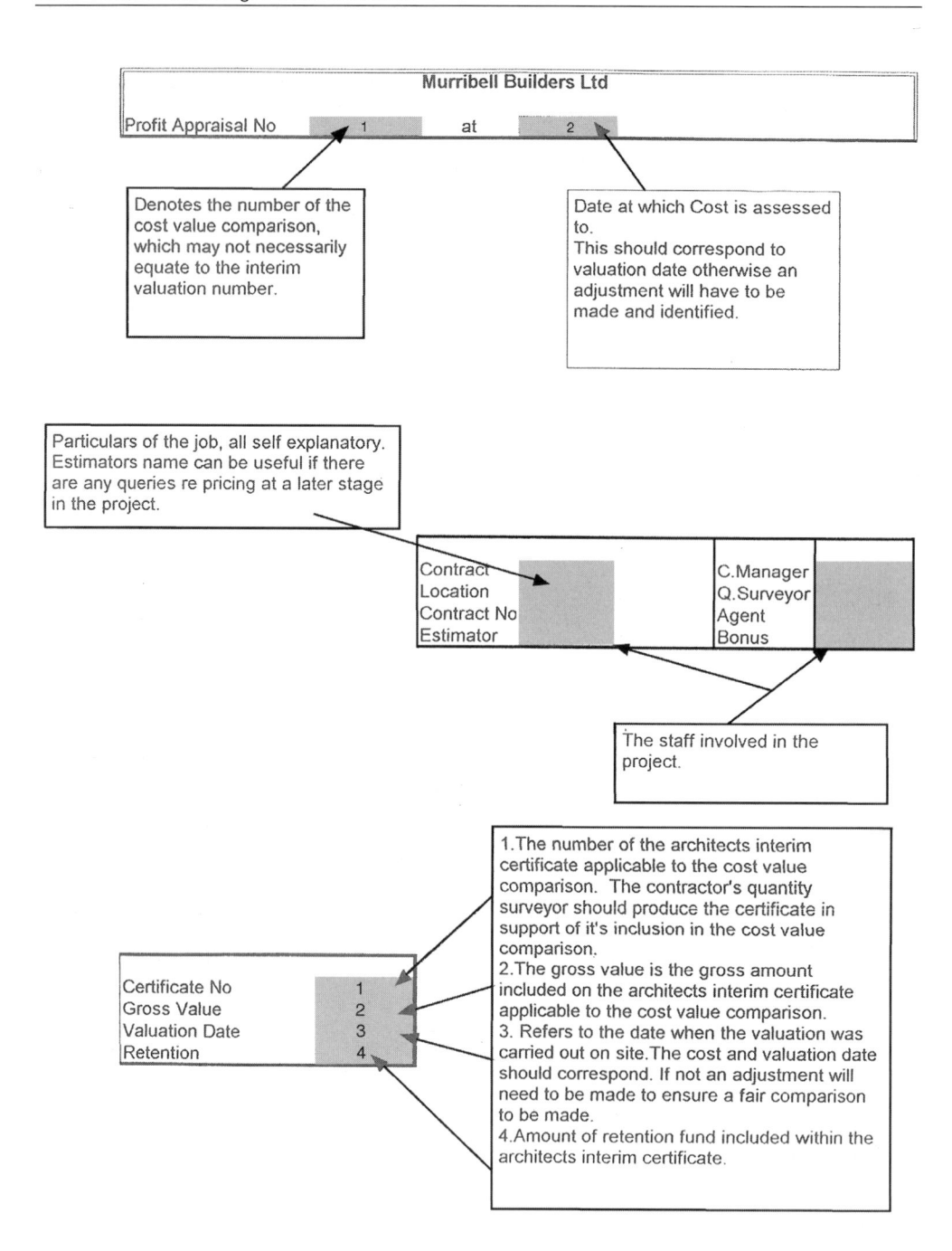

Murribell Builders Ltd

Profit Appraisal No ⬚ 1 ⬚ at ⬚ 2 ⬚

- Denotes the number of the cost value comparison, which may not necessarily equate to the interim valuation number. *(annotation for 1)*
- Date at which Cost is assessed to. This should correspond to valuation date otherwise an adjustment will have to be made and identified. *(annotation for 2)*

Particulars of the job, all self explanatory. Estimators name can be useful if there are any queries re pricing at a later stage in the project.

Contract		C.Manager	
Location		Q.Surveyor	
Contract No		Agent	
Estimator		Bonus	

The staff involved in the project.

Certificate No	1
Gross Value	2
Valuation Date	3
Retention	4

1. The number of the architects interim certificate applicable to the cost value comparison. The contractor's quantity surveyor should produce the certificate in support of it's inclusion in the cost value comparison.
2. The gross value is the gross amount included on the architects interim certificate applicable to the cost value comparison.
3. Refers to the date when the valuation was carried out on site. The cost and valuation date should correspond. If not an adjustment will need to be made to ensure a fair comparison to be made.
4. Amount of retention fund included within the architects interim certificate.

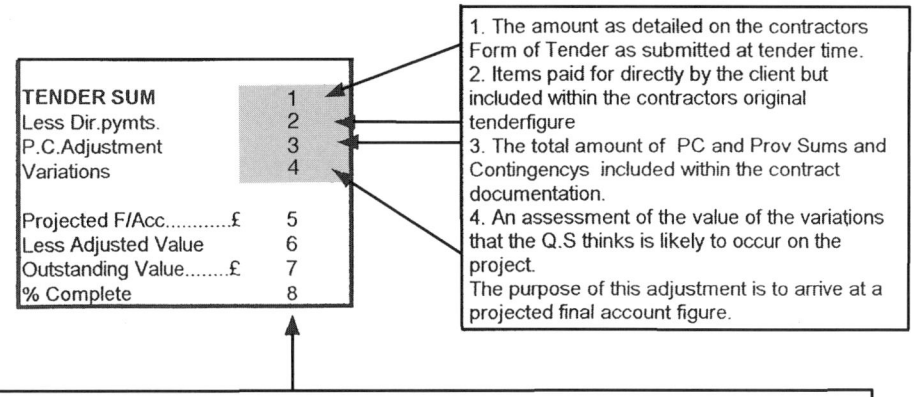

Appendix 2: Cost value comparison

TENDER SUM	1
Less Dir.pymts.	2
P.C.Adjustment	3
Variations	4
Projected F/Acc............£	5
Less Adjusted Value	6
Outstanding Value........£	7
% Complete	8

1. The amount as detailed on the contractors Form of Tender as submitted at tender time.
2. Items paid for directly by the client but included within the contractors original tenderfigure
3. The total amount of PC and Prov Sums and Contingencys included within the contract documentation.
4. An assessment of the value of the variations that the Q.S thinks is likely to occur on the project. The purpose of this adjustment is to arrive at a projected final account figure.

5. The Projected F/Acc figure is the QS's assessment at the time of the production of the cost value comparison of the likely final contract figure. This figure should be reviewed monthly as circumstances change throughout the currency of the project.
6. This figure is the amount of the architects interim valuation applicable to the CVC, adjusted to account for any under or over valuation.
7. The balance of the projected final account left to expend on the project (projected final account less current adjusted value).
8. Simply the outstanding value shown as a percentage how much of the job is complete this can be compared.

1. The number of weeks extensionof time applied for.
2. The actual amount of weeks that have been certified / awarded.
3. The number of weeks that the Q.S actually thinks will be awarded. On their best judgement of the facts they presented to the client.

			Contract	Prelims
Contract Commencement Date			4	5
Contract Completion Date			6	7
Contract Weeks Ahead/(behind)			8	9
Revised Completion Date			10	11
			Contract	Prelims
Ext.App.For	1	Weeks	12	13
Ext.Granted	2	Weeks	14	15
Ant.F/Awd	3	Weeks	Spare	Spare

4. & 5. The contract will start on a set date and the prelims will start on a set date and this is what is shown here. The two don't necessary have to start on the same date.
6. & 7. This is exactly the same as above but it shows when the contract and the prelims will be complete.
8. & 9. The amount of weeks that the project is ahead/behind for both the contract and the prelims.
10. & 11. As a result of what is shown in boxes 8 & 9 the revised completion date will be calculated and shown here.
12.13.14.15.The progress of the project shown as a percentage complete / fraction complete

170 Commercial management in construction

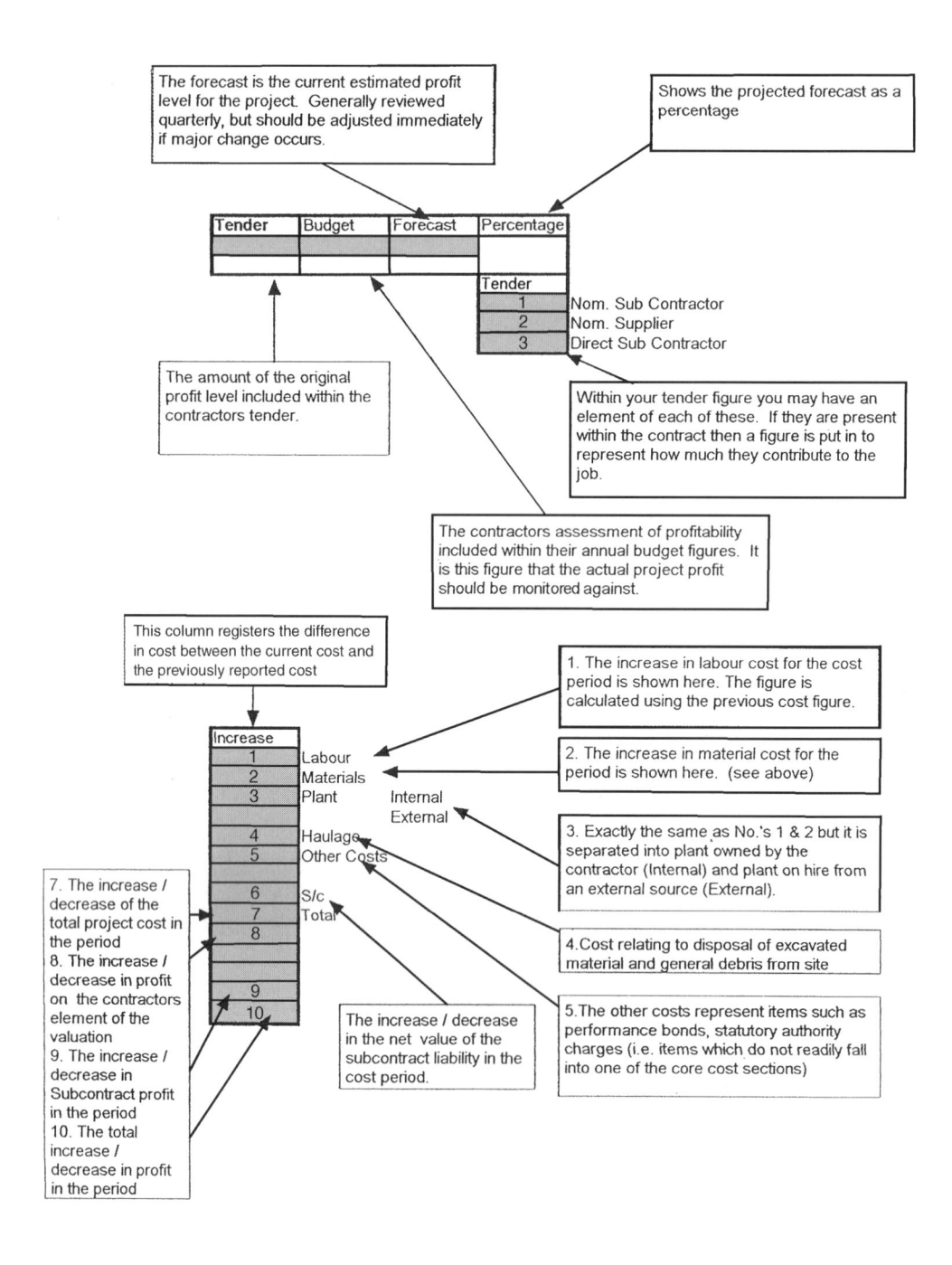

The forecast is the current estimated profit level for the project. Generally reviewed quarterly, but should be adjusted immediately if major change occurs.

Shows the projected forecast as a percentage

Tender	Budget	Forecast	Percentage

Tender	
1	Nom. Sub Contractor
2	Nom. Supplier
3	Direct Sub Contractor

The amount of the original profit level included within the contractors tender.

Within your tender figure you may have an element of each of these. If they are present within the contract then a figure is put in to represent how much they contribute to the job.

The contractors assessment of profitability included within their annual budget figures. It is this figure that the actual project profit should be monitored against.

This column registers the difference in cost between the current cost and the previously reported cost

Increase	
1	Labour
2	Materials
3	Plant — Internal / External
4	Haulage
5	Other Costs
6	S/c
7	Total
8	
9	
10	

1. The increase in labour cost for the cost period is shown here. The figure is calculated using the previous cost figure.

2. The increase in material cost for the period is shown here. (see above)

3. Exactly the same as No.'s 1 & 2 but it is separated into plant owned by the contractor (Internal) and plant on hire from an external source (External).

4. Cost relating to disposal of excavated material and general debris from site

5. The other costs represent items such as performance bonds, statutory authority charges (i.e. items which do not readily fall into one of the core cost sections)

The increase / decrease in the net value of the subcontract liability in the cost period.

7. The increase / decrease of the total project cost in the period
8. The increase / decrease in profit on the contractors element of the valuation
9. The increase / decrease in Subcontract profit in the period
10. The total increase / decrease in profit in the period

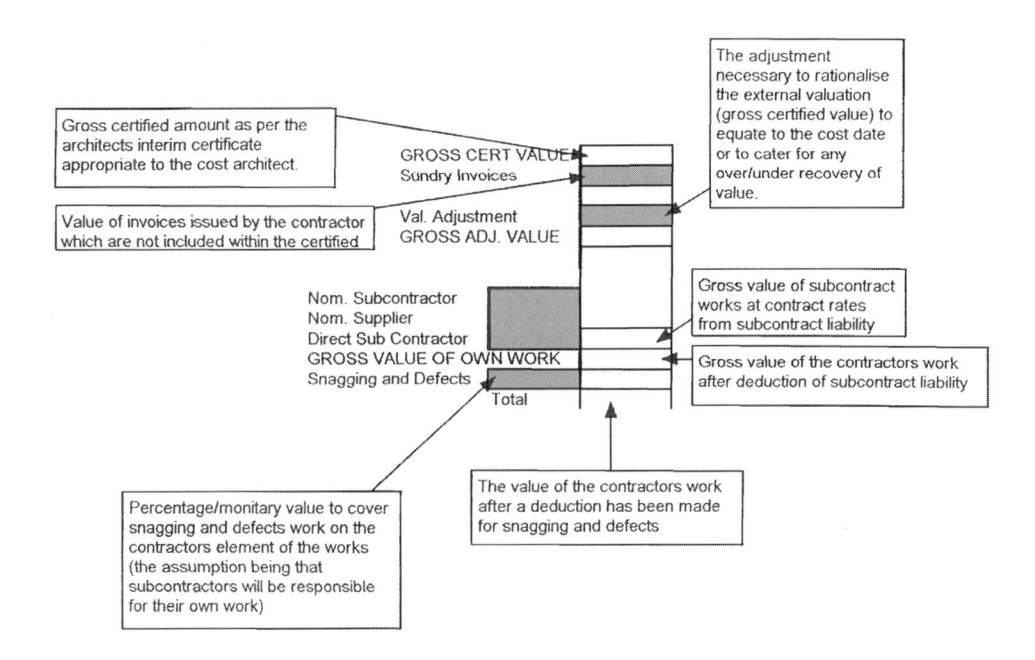

Appendix 2: Cost value comparison

Gross certified amount as per the architects interim certificate appropriate to the cost architect.

GROSS CERT VALUE
Sundry Invoices

The adjustment necessary to rationalise the external valuation (gross certified value) to equate to the cost date or to cater for any over/under recovery of value.

Value of invoices issued by the contractor which are not included within the certified

Val. Adjustment
GROSS ADJ. VALUE

Nom. Subcontractor
Nom. Supplier
Direct Sub Contractor
GROSS VALUE OF OWN WORK
Snagging and Defects
Total

Gross value of subcontract works at contract rates from subcontract liability

Gross value of the contractors work after deduction of subcontract liability

Percentage/monitary value to cover snagging and defects work on the contractors element of the works (the assumption being that subcontractors will be responsible for their own work)

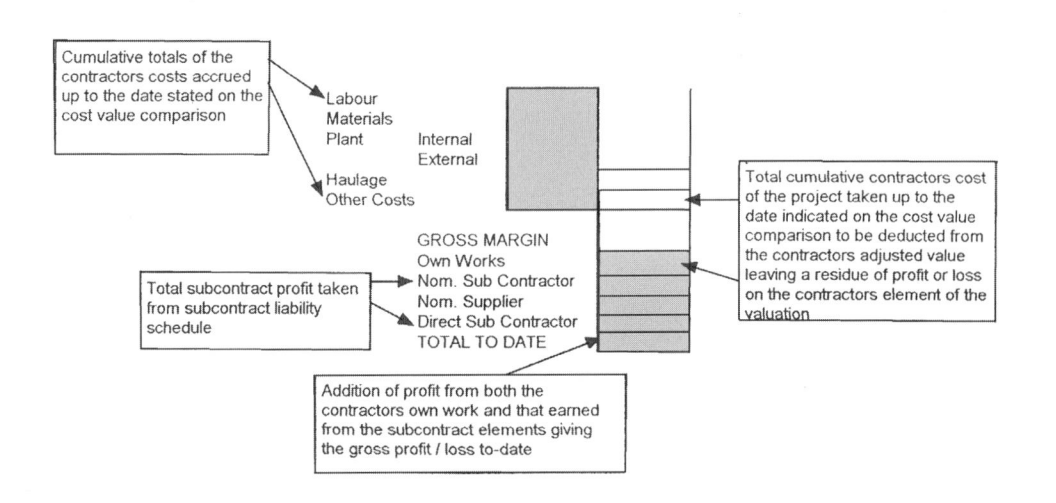

The value of the contractors work after a deduction has been made for snagging and defects

Cumulative totals of the contractors costs accrued up to the date stated on the cost value comparison

Labour
Materials
Plant   Internal
            External
Haulage
Other Costs

GROSS MARGIN
Own Works
Nom. Sub Contractor
Nom. Supplier
Direct Sub Contractor
TOTAL TO DATE

Total subcontract profit taken from subcontract liability schedule

Total cumulative contractors cost of the project taken up to the date indicated on the cost value comparison to be deducted from the contractors adjusted value leaving a residue of profit or loss on the contractors element of the valuation

Addition of profit from both the contractors own work and that earned from the subcontract elements giving the gross profit / loss to-date

C.V.C (Cost Value Comparision) Explained.

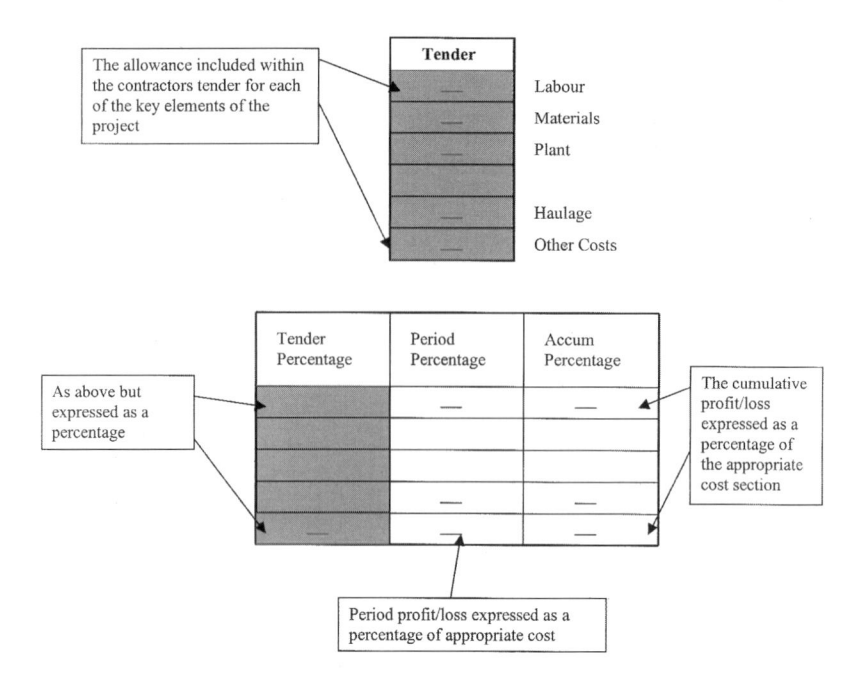

Tender

Labour
Materials
Plant
Haulage
Other Costs

The allowance included within the contractors tender for each of the key elements of the project

Tender Percentage	Period Percentage	Accum Percentage
	—	—
	—	—
—		—

As above but expressed as a percentage

The cumulative profit/loss expressed as a percentage of the appropriate cost section

Period profit/loss expressed as a percentage of appropriate cost

Bibliography

Aqua Group (1996) *Contract Administration for the Building Team*. Blackwell Scientific Publications.
Barrett, F.R. (1981) *Cost Value Reconciliation*. Chartered Institute of Building.
Brandon, P.S. (ed.) (1992) *Quantity Surveying Techniques New Direction*. Blackwell Scientific Publications.
Chartered Institute of Building (1983) *Code of Estimating Practice* (Fifth Edition). Chartered Institute of Building.
Chartered Institute of Building (1998) *Education and Professional Development Manual Section 2*.
Chartered Institute of Building/DfEE (1997) 'Sharing best practice in employment related building management education'. *A Higher Education Discipline Network*, October.
Cottrell, G.P. (1978) *The Builder's Quantity Surveyor*. CIOB Surveying Information Service No. 1.
Dent, C. (1970) *Quantity Surveying*. Oxford University Press.
Department of the Environment, Transport and the Regions (1999) *A Working Definition of Local Authority Partnerships*. Newchurch.
Department of the Environment, Transport and the Regions (1999) *New Partnerships with the Private Sector*. Hilary Armstrong.
Department of Trade and Industry (1998) *Partnering for Profit*. Stationery Office.
Dickason, I. (1982) *JCT 80 and the Builder*. Chartered Institutue of Building.
Franks, J. (1991) *Building Contract Administration and Practice*. B.T. Batsford.
Gill, J. and Johnson, P. (1997) *Research Methods for Managers* (Second Edition). Paul Chapman.
GTI (1998) *GTI Quantity Surveying* (Edition 11 1999).
Inland Revenue (1998) *Construction Industry Scheme*. External Communications Unit of the Inland Revenue, December.
Institute of Chartered Accountants (1992) *Statement of Standard Accounting Practice No. 9*.
Jones, G.P. (1980) *A New Approach to the 1980 Standard Form of Building Contract*. Construction Press.
Keating, D. (1978) *Building Contracts*. Sweet and Maxwell.
Pilcher, R. (1985) *Project Cost Control in Construction*. Collins.
QS Think Tank (1998) *The Challenge of Change*. RICS.
Ramus, J.W. (1989) *Contract Practice for Quantity Surveyors* (Second Edition). Newnes.
RICS, Quantity Surveying Division (1995) *Chartered Quantity Surveyors in a Changing World: An Agenda for Change*.
Seeley, I.H. (1997) *Quantity Surveying Practice* (Second Edition). Macmillan.
Seeley, I.H. (1998) *Building Quantities Explained* (Fourth Edition). Macmillan.
Wainwright, W.H. (1967) *Variation and Final Account Procedure*. Hutchinson Technical Education.
Wallace, I.N.D. (1995) *Hudson's Building and Engineering Contracts*. Sweet and Maxwell.
Willis, A. and Tench, W. (1998) *Willis's Elements of Quantity Surveying* (Ninth Edition). Blackwell Science.
Willis, C.J., Ashworth, A. and Willis A. (1994) *Practice and Procedure for the Quantity Surveyor* (Tenth Edition). Blackwell.

Glossary of Terms

Anticipated tender conversion	Estimate on unknown projects anticipated to be obtained throughout the period of the budet or forecast
Budget	Annual estimate of projected company performance which, once established, does not alter and is used as a tool against which to monitor actual performance
CIOB	Chartered Institute of Building
CIS	Construction industry (tax) scheme
Cost value comparison (CVC)	Format used to compare actual cost against valuation on a project-specific basis
DETR	Department of the Environment, Transport and the Regions
Domestic subcontractor	Subcontractor selected solely by the contractor
Extension of time	Assessment of time by the architect added to the contract 'Date for Completion'
Final account	Final agreed value of any project
Financial accounts	Statement of company performance at the end of a company's financial year
Forecast	Similar to the budget, but is reviewed and revised on a regular basis, reflecting the company's position at a given time
Gross margin	Total profit achieved before the deduction of the total overhead cost
IFC	Intermediate form of contract
Interim valuation	Regular assessment of valuation of the works carried out to allow the contractor regular payments on account
JCT	Joint Contracts Tribunal
KPI	Key performance indicator
Liquidated and ascertained damages	Predetermined amount withheld from payments to the contractor following the contractor's default in completing the works in accordance with the contract provisions
Management accounts	Review of company performance based on current assessments of valuation, cost and overhead costs

Materials on site	Schedule of materials held on site before fixing, valued and included within interim valuations
Named subcontractor	Subcontractor selected by the architect without the strict provisions and rules of nominated subcontractors
Net cost of production (NCP)	Relates to the actual direct cost involved in the completion of any project, including all of the contractor's labour, plant, material and ancillary charges together with all subcontracting costs
Nominated subcontractor	Subcontractor selected by the architect under special provisions
Overheads	Generally relate to the total cost of the contractor's office establishment, including the cost of all staff and their associated charges, except those charged directly to projects, office rent and rates, insurance, general office running costs, heating, lighting, stationery, etc.
PC	Prime cost
Profit	Residue left after all costs are accrued and finalised and deducted from the agreed total value, on either a project-specific or a company basis
Retention	Amount withheld from an interim valuation in accordance with the conditions of contract
RIBA	Royal Institute of British Architects
RICS	Royal Institution of Chartered Surveyors
Stage payment schedule	Analysis of contract works tabulated and valued from the bills of quantities in work sections used for interim valuation purposes
Subcontract comparison	Analysis of a selection of subcontractors' work packages compared with original bill of quantity allowances
Subcontract liability	Assessment of a subcontractor's value included within an interim valuation or final account compared with the value that the contractor will be paid for the same elements of work

Index

Activity schedule 18
Actual results vs budget 13
Addendum bill 26
Annual budget 6
Appendix to conditions 93
Approximate quantity 31
Architects 17
   contractor failure 46
   interim certificates 21, 42–43, 60, 97
   making good defects and additional
      works 101
   practical completion 99
   role of 2–3
   subcontractor services 57, 58, 59, 61, 62,
      64–65, 66
   timescales 61, 105
   see also Royal Institute of British Architects
Architect's instructions 19, 28–41, 51, 52, 85,
   107, 120
Architect's interim valuation certificate 80–81

Bill of quantities 24, 25
Bonuses and other incentive payments 36
Brick and blockwork 24
Budget, definition of 6, 174
Building Cost Information Service 37
Building Employers Confederation 31, 40

Cash flow 20
Certificate of Completion of Making Good
   Defects 17, 56, 100–103
Certificate of Practical Completion 17,
   98–100
Chartered Institute of Building (CIOB) 3–4
Client satisfaction 113
Company budget 6
Compensation events 115
Completed projects 8
Condition precedents 93
Conditions of contract 93
Conquest valuation package 25
Construction Industry Scheme (CIS) 69, 77,
   174

Construction time 114
Contrack Benchmark 114
Contract analysis schedule 53, 54
Contract commencement 93–95
Contract review meetings 117–118
Contractor core costs 88–92
Cost value comparison 8, 13, 79, 81, 88,
   167, 174
   schedule 80
Credit arrangements with suppliers 21
Current projects 6

Date of possession 20
Daywork 33, 39
   voucher 34
Defects 114
Defects Liability Period 17, 56
Delay schedule 106
Department of the Environment, Transport
   and the Regions (DETR) 174
Direct payment procedure 43
Domestic subcontractors 66–67,
   76, 174
Drainage and service installation
   24–25
Drivers of change 111

Egan's construction task force 111
Estimating packages 26
Excel spreadsheets 11
Expansion programmes 7
Extension of time 103–105, 174

Final accounting 51–56, 174
   statement 53
Final certificate 107–108
Final contract adjustments 56
Finance director, role of 6
Financial accounts 174
Fluctuations 19, 55
Forecasts 13–16, 174
   revisions 13
   summary 11, 12

Gross adjusted valuation 82–83
Gross certified value 82, 83
Gross margin 13, 89, 174

Head office charges 36
Health and Safety Commission 116
Hourly rates 35

Incidental costs 36–37
Institute of Chartered Accountants 78
Insurance costs 37
Interim certificates 17, 18, 97–98
　deductions 47
Interim valuation 120–138, 174
　elements of 18–21
　information 80
　method of production 21
　preparing 19–20
　timing of 17
Intermediate form of contract (IFC) 174

Joint Contracts Tribunal 17, 174

Key performance indicators 113–116, 174

Labour 34–35
Labour-only subcontractors 67–69
Letters of Intent 95–97
Liquidated and Ascertained Damages
　46–47, 174
Look-ahead schedule 113
Loss and expense 105–107
Lotus 1-2-3 spreadsheets 11

Management account summary
　15
Management accounts 174
Material analysis 8
Materials and goods 36
Materials off site 19, 45
Materials on site 43–44, 175
Measured works 19

Named subcontractors 62–64, 175
National Federation of Builders,
　history of 5
National Insurance contributions 37
Net cost of production 6, 13, 79, 175
Net profit 13
Nominated subcontractors 57–62, 175
　valuation of 42
Nominated supplies, valuation of 42

Overheads 13, 36–37, 39, 175
　assessment of 6
Overtime 36

Partnering 2, 109–110
Payment 17, 47
Performance monitoring
　113–114
Period of Interim Certificates 17
Photographs 49
Plant 36
Possible projects 7
Postcontract procedure 65–66
Precontract procedure 64–65
Predictability 114
Preliminaries 19, 26–28,
　120, 153
　draw sheet 27, 29
Price, subcontractor selection 69
Prime cost 33
　sums 19
Prime Cost of Daywork 31, 33
Probable projects 6
Productivity 114
Profit 13, 36–37, 39, 175
　assessment of 6, 11
Profitability 114
Programming of works 71
Project analysis 6
Project forecast 8, 9
Provisional assesment form 50
Provisional items 19, 41–42
Provisional quantities 19, 84

Quantity surveyor, role of 6, 7, 8, 11, 17,
　19–20, 21, 23, 24, 25, 30, 32, 39, 40, 41,
　42–43, 44, 46, 47, 49, 51, 52, 74, 77, 82,
　83–84, 85, 88, 90, 94, 97, 99, 101, 103,
　104–105, 119, 120, 139–140, 153

Refurbishment projects 40–46
Remeasurement 41–42
　schedules 24
Retention 18, 45–46, 175
Risk 11
Risk register 13, 14
Royal Institute of British Architects
　(RIBA), history of 4–5
Royal Institution of Chartered
　Surveyors 31, 40
Royal Institution of Chartered Surveyors
　(RICS), history of 2–3

Index

Safety 114
'S'-curves 11
Sick pay 37
Site
   staff 36
   valuation notes 24
   visits 49–51
Snagging and defects 85–88
Specification changes 70
Spreadsheets 11
Staffing levels 7
Stage payments
   analysis 23
   schedule 22, 24, 175
Standard Hourly Base Rates 37–39
Standard Method of Measurement of Building
   Works 31, 40, 41
Statement of Standard Accounting
   Practice No. 9 (SSAP9) 1, 78
Subcontract 71
   comparison 10, 68, 175
   liabilities 85, 153, 175
   order, placing 73–76
   package 69
   payments 76–77
      schedule 75
   registration cards 77–77
   tax certificates 77–77
Subsistence 36

Tax invoice
Team
   formation 112–113
   meetings 117
Tender
   analysis 26
   conversions, anticipated 6, 7–13, 174
   qualifications 70–71
   success ratios 7
   sum analysis 89
Time sheets 50
Tool allowances 37
Travelling 36

Unfixed materials on site 19

Valuation adjustment 83–85
Valuation packages, computerised 25
Value added tax (VAT) 47–49
Variations 28–41
Variations and provisional sums 30

Weather 36
Weekly earnings 38
Work in progress 6
Working hours 30, 37
Working space 30

Zero accidents 116